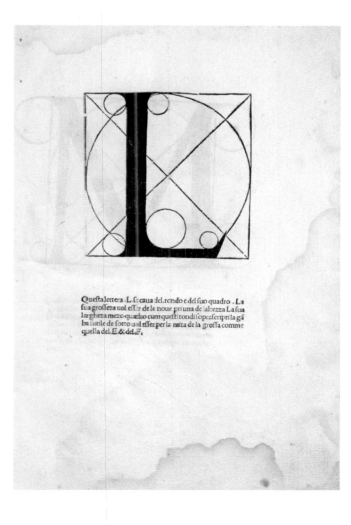

Questa lettera .L. se caua del rondo e del suo quadro . La
sua grosseza uol esser de le noue prima de la lezza La sua
larghexa mezo quadro cum questi rondi soprascripti la gã
ba tutile de sotto uol esser per la mtta de la grossa comme
quella del. E. & del. F.

LEONARDO'S UNIVERSE

The Renaissance World of Leonardo da Vinci

BULENT ATALAY AND KEITH WAMSLEY

NATIONAL GEOGRAPHIC

WASHINGTON, D.C.

"disscepolo della sperientia"

CONTENTS

*Opposite: Leonardo, "Self-portrait," ca 1510-13. Biblioteca Reale, Turin. Leonardo signed off
in his notebooks as "disscepolo della sperienza," which would be rendered in modern Italian as* discepolo
dell'esperienza *(disciple of experience). Page 1: A letter in a font introduced in mathematician Luca Pacioli's
treatise* De Divina Proportione. *Leonardo illustrated the volume. Pages 2-3: The bell tower of the Duomo,
Florence, reflected in a puddle of water*

> *"Leonardo da Vinci was like a man who awoke too early in the darkness, while the others were all still asleep."*
>
> —SIGMUND FREUD

PROLOGUE

The name of Leonardo da Vinci and the term Renaissance, the designation historians have given to the period during which Leonardo lived, have become inextricably woven together. Who has not heard Leonardo called the quintessential "Renaissance man"? And who can pretend to even a casual knowledge of the European Renaissance without knowing something of the accomplishments of Leonardo? The man and the time have maintained their hold on us.

The European Renaissance—especially its beginnings in 15th century Italy—is widely recognized for its great art and architecture, and this book, with the matchless graphics and photography of the National Geographic Society, brings the full splendor of the artists and craftsmen of that age to its pages. In addition to the great art of the past, there are photographs of modern Italy, many dominated by monuments of the Renaissance. They serve to connect the present with a time and place in which many of the foundations of our modern civilization were laid. In the globalizing world of today, wherein a multitude of societies and cultures appear to be reducing themselves into an homogenized common civilization, there is no more powerful agent of that transformation than Western science and technology.

The seeds of that science and technology, which germinated and flowered in the Scientific Revolution of the 17th century, were planted in the Italian Renaissance of the 15th century, and no one sowed more of those seeds than Leonardo. His curiosity about the universe in which he found himself made him typical of the new intellectual outlook of his time, but no one else of the time was, as art critic Kenneth Clark put it, more "relentlessly curious." He filled his voluminous notebooks

Pages 6-7: Morning fog envelops Leonardo's native Tuscany, a rich agricultural region since ancient times. Pages 8-9: The heart of Florence in Leonardo's time was the Duomo complex, featuring Brunelleschi's dome and Giotto's bell tower; the tower on the right belongs to the Palazzo Vecchio. Pages 10-11: The Duomo's floor is a sea of geometric and floral designs.

with original scientific observations, speculations, and hypotheses, most of which would be borne out and supported by later independent researchers in the coming centuries. And he sketched designs for countless engines and machines, many of which would later make an actual appearance in the world. Though in his own lifetime not much was known by others of his inventions and scientific endeavors—for Leonardo was often secretive and never published any of this work—his science partakes fully in the spirit of the age.

The Italian Renaissance was intellectually expansive, reaching back to recover the art, philosophy, and literature of the ancients, but also taking the first steps into a new kind of human future. The first comprehensive designs for that future are found in the notebooks of Leonardo, and the very best of them are displayed in these pages. When we encounter them today across the span of 500 years, they appear even more marvelous than they must have seemed in their own time. It is that encounter that causes each succeeding generation to regard Leonardo as a prophet of science and technology, the first and most successful futurologist in history.

Leonardo the early Renaissance scientist was also a supreme artist, though his artistic output was meager, especially by the standards of his time. He brought the same energy to his art that he brought to his science and invention. He was always pushing against the boundaries, experimenting with new techniques, conceptualizing spectacular productions, again in keeping with the spirit of his times. But the mind of Leonardo was as restless as it was relentless. With the exception of a few of his scientific interests, such as flight and human anatomy, nothing held his attention for long. He would frequently suspend even these interests, only to return to them later with renewed vigor. Some of his best paintings remain unfinished, abandoned under the remorseless crush of too many things to be done, too many investigations to pursue, too many questions to ask and answer. Often he simply could not take the time to linger over the execution of a single work of art once it had been conceived, visualized, and planned. Once the mental work was done, despite his immense talent, the manual labor was beneath him. More important things beckoned. Being a Renaissance man was demanding, even in the Renaissance.

The more one is acquainted with the intellect of this man who lived half a millennium ago, the greater the respect one has both for him as an individual and for the time and place that produced him. He was a great protoscientist before the Scientific Revolution and a great artist. The two roles were never distinct. They shared the soul of Leonardo, just as great art and the new liberating intellectual sensibility share what is at the heart of the Renaissance. In that unique time, so important to the long development of our civilization, he was a uniquely positive voice, one who represents humanity at its very best.

Chapter One

"There was an infinite grace in all his actions; and so great was his genius, and its growth, that to whatever difficulties he turned his mind, he solved them with ease."

—GIORGIO VASARI

THE ITALIAN RENAISSANCE AND
THE RENAISSANCE MAN

1296 Construction begins on Florence's Duomo.

ca 1300 Humanism arises in Italy.

1332 Giotto designs bell tower for Florence's Duomo.

1348 Black Death spreads to Italy.

1353 Boccaccio, father of Italian prose, completes the *Decameron*.

1394 Milanese invade Florence.

1397 The Medici establish a banking house in Florence.

1401-02 Ghiberti wins competition for second set of baptistery doors.

1419 Brunelleschi designs dome for Florence's cathedral. He formulates theory of linear perspective.

1425-1452 Ghiberti constructs Doors of Paradise.

ca 1435 Alberti writes first treatise on theory of painting.

1436 Brunelleschi completes the dome for Florence's Duomo.

ca 1444 Donatello casts "David," the first freestanding male nude since antiquity.

1450-54 Gutenberg prepares first published book by moveable type.

1453 Constantinople falls to Ottoman Turks.

The life of Leonardo da Vinci spanned 67 years of the European Renaissance, a period of discovery and expansion as well as conflict and warfare. In Leonardo's time, his native Italy enjoyed unprecedented prosperity and preeminence—becoming the nexus of one of the greatest periods of creativity in the history of civilization. Leonardo's birth in 1452 very nearly coincided with Johannes Gutenberg's introduction of printing by movable type in Mainz, Germany; Leonardo was barely a year old when Constantinople, the capital of the Byzantine Empire, fell to the Ottoman Turks on May 29, 1453; and he was 40 years old when Christopher Columbus landed in the New World in 1492.

Renaissance Sensibilities

This shift in power in Constantinople led to the emigration of Byzantine scholars, with their valuable manuscripts from antiquity and their knowledge of ancient Greek, to Italy and the West—an event often heralded as marking the end of the Middle Ages. In Western Europe these scholars were embraced by a group known as the humanists, who, in reaction to the narrow, theologically focused learning of the time, advocated a new approach. Based on the classical literature and learning of ancient Greece and Rome, their aims, they believed, would release the best and broadest energies of the human intellect; they had been at work since the mid-1300s searching dusty libraries all over Europe for Latin manuscripts from the almost forgotten Roman past. This renaissance, or "rebirth," of the classical past eventually gave its name to the era. The humanists brought with them a new sensibility about the destiny of humanity, the dignity of the individual, and the goals of the artist—and, especially after the introduction of the printed book, this new humanism spread rapidly.

The new spirit of the time had arisen gradually over the previous three centuries as changes in medieval society accumulated, deepened, and finally achieved an irresistible momentum. It was not by chance that these energies first converged in

Opposite: Francesco Melzi, "Leonardo in Profile," ca 1510. Royal Collection, Windsor Castle, U.K.
Previous pages: View of Florence from Piazza Michelangelo, with the Arno River in the foreground

LEONARDO

LEONARDO'S EARLIEST WORK

Leonardo's ability to depict depth and his legendary handling of light and shadow are evident in a sketch of the Arno Valley (right). Leonardo acquired his wonder and passion for nature in his childhood in the verdant hills of his native Tuscany. This, his earliest known landscape, a pen-and-ink study dated August 5, 1473, when he was 21 years old, depicts his beloved Arno Valley seen from a hilltop. The scene is rich with artistic devices—impeccable perspective, precise shading, and an eye for detail. Already present in this early drawing are the degradation in the light with distance, as seen through the eyes of a physicist, and the rock formations—which will later form the backdrop in several of his greatest paintings—reflecting the interests of a geologist. On display, even at this young age, is his lifelong practice of bringing together science and art. Probably drawn in Florence, it represents an amalgam of his childhood memories.

Italy, the crossroads of the Mediterranean world. Even after the collapse of ancient Roman power and authority following the invasion of the barbarian hordes from the north in the fifth to eighth centuries, trade, almost at a halt elsewhere, continued to flourish in Italy. Italian cities such as Genoa and Venice thrived on economic exchange with the Byzantine East and the Islamic lands. Through this trade, urban life survived in Italy to an extent unknown in the rest of western Europe. Italy became a land of city-states where commerce flourished and adventurers and explorers sought to advance their commercial interests. Marco Polo, from the city of Venice, journeyed to the Far East in the 13th century and brought back detailed

descriptions of China and Japan, and the Genoese sailor Columbus, in search of new trade routes to the East and in the employ of the Spanish, led a European expedition that profoundly changed the fates of both the New World and the Old.

Trade made the cities of Italy prosperous. By 1400 the wealth of the merchants, manufacturers, and bankers of many towns and cities was beginning to surpass that of the traditional landed aristocracy. Rich and powerful families produced men of affairs who took control of the city-states, or communes, which included the cities and the surrounding countryside and villages under their political control. These great families, such as the Este of Ferrara, the Gonzaga of Mantua, and, most notably, the Medici of Florence, began to use their fortunes to enhance public spaces with art and architecture to reflect their cities' power and the civic pride of the leading families. Cathedrals, churches, monasteries, and nunneries were improved and embellished. Magnificent libraries were assembled; new schools were endowed. The new humanist scholars were welcomed at court and encouraged in their pursuit of recovering the learning of antiquity—and architects, artists, and sculptors were kept busy in their workshops producing works of startling freshness and originality. Kings, dukes, popes, doges, and maestros competed to attract the greatest scholars, architects, and artists to their own cities and courts.

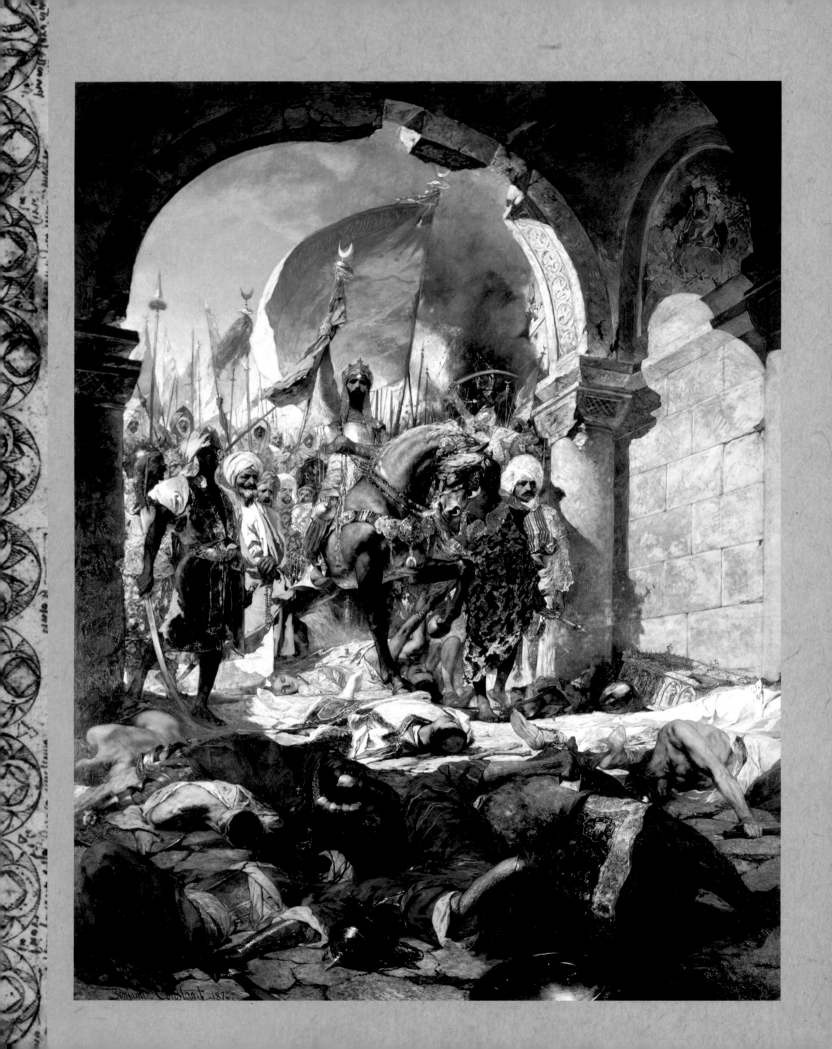

THE FALL OF CONSTANTINOPLE

> *"On the 24th day of 1453, ... Mahomet Bey, son of Murat the Turk, came himself to the walls of Constantinople to begin the general assault which gained him the city."*
> —NICOLO BARBARO

The land occupied by modern Turkey's largest city, Istanbul, has been continuously settled for more than 3,000 years. In the eighth century B.C. the colonizer of cities, Byzans, on advice from the Oracle of Delphi, founded the Greek city of Byzantium on the western banks of the Bosporus. Some 1,100 years later, in A.D. 324, Constantine, who had made Christianity the official religion of the Roman Empire, chose Byzantium as his new capital city. After Constantine's death in A.D. 337, it was renamed Constantinople.

The empire began to separate into two distinct political divisions, the East eventually becoming the Byzantine Empire, and the West finally disappearing as a political entity after being ravaged by barbarian invaders from the north. Rome survived as the headquarters of western Christianity. Differences in language and doctrine had led to the great schism of 1054, the church in the Latin West remaining under the authority of the pope in Rome, and the church of the Greek-speaking East separating under the authority of their local bishops.

Opposite: Jean-Joseph Benjamin-Constant, "Entry of Mehmet II into Constantinople," 1876. Musée des Augustins, Toulouse

Above: Built in the 12th century, the Church of St. Savior in Cora, Istanbul, now the Kariye Museum, is prominent for its mosaics and frescoes.

As the capital of the Byzantine Empire, Constantinople survived countless sieges before it fell to the friendly forces of the Fourth Crusade in 1204. This army had set out to retake the Holy Lands from the Muslims but got sidetracked, settling in Constantinople instead. The sack of the city led to the seizure of untold classical treasures, including the four gilded horses now at St. Mark's Cathedral in Venice. This occupation would last only half a century, but Constantinople would never recover. In 1347 its fortunes would sink even lower with the arrival of the bubonic plague.

Meanwhile, the Ottoman Turkish State, with its capital in nearby Bursa (Roman Prusa) began to mount a sustained siege around a shrinking Byzantine Empire. Finally, in 1453, the Ottomans, under their leader Sultan Mehmet I, "Mehmet the Conqueror," breached the walls of the city, and defeated the forces of Constantine XI, the last Byzantine emperor. The attackers vastly outnumbered the defenders and possessed heavier artillery, but they also had to break through fortified walls described as the strongest anywhere in the world. The fall of Constantinople signaled the end of the Byzantine state itself, and is often used as a convenient marker for the end of the Middle Ages.

Preceding the final assault on Constantinople, many Byzantine scholars had emigrated to Italy at the invitation of the Italian

humanists and taken jobs as teachers and translators. Among them were Theodore Gaza, John Argyropoulos, and the most influential of all, Demetrius Chalcondyles. These scholars brought with them a wealth of classical manuscripts, as well as the knowledge needed to access them. The rich infusion of scholarship is frequently cited as one of the most significant contributions to the Italian Renaissance.

Below: Gentile Bellini, "Portrait of Mehmet II, 'the Conqueror,'" 1480. National Gallery, London

Italian School, "Carta della Catena," showing a panorama of Florence, 1490. The city walls of Florence were still intact in Leonardo's day.

The prosperous and powerful city-state of Florence in the mid-1400s was the backdrop for the early artistic career of Leonardo. By the time Leonardo arrived in Florence as a teenager in the 1460s, it was already at the forefront of the arts and of humanist learning in Italy. Most of the important Renaissance artists of the previous generation had worked in Florence, and many of the rising generation, including Leonardo, made it their artistic school and home.

Mid-15th-century Florence had a population of 100,000—small compared with its neighbors—but the inhabitants prided themselves on their fiercely independent spirit. Florentines identified as much with their distant forebears, the

Etruscans, who had occupied the same region, as with the ancient Romans. Florence adopted as its emblem the Old Testament hero David—known for his indefatigable spirit and for slaying the giant Goliath. The skyline of Florence, then as now, was dominated by the Duomo di Santa Maria del Fiore, its largest and most important building. The Duomo, begun in 1296 and built according to the original conception of medieval architect Arnolfo di Cambio, ultimately took more than 170 years to build—not unusual for medieval and Renaissance cathedrals. As of 1400, however, the Duomo had only an uninspiring and undistinguished wooden dome. At the beginning of the 15th century, Florentines determined that the cathedral of

Following pages: The polychromic—green, white, and red—revetment, or facing, of Florence's Santa Maria del Fiore was designed to match the earlier existing baptistery.

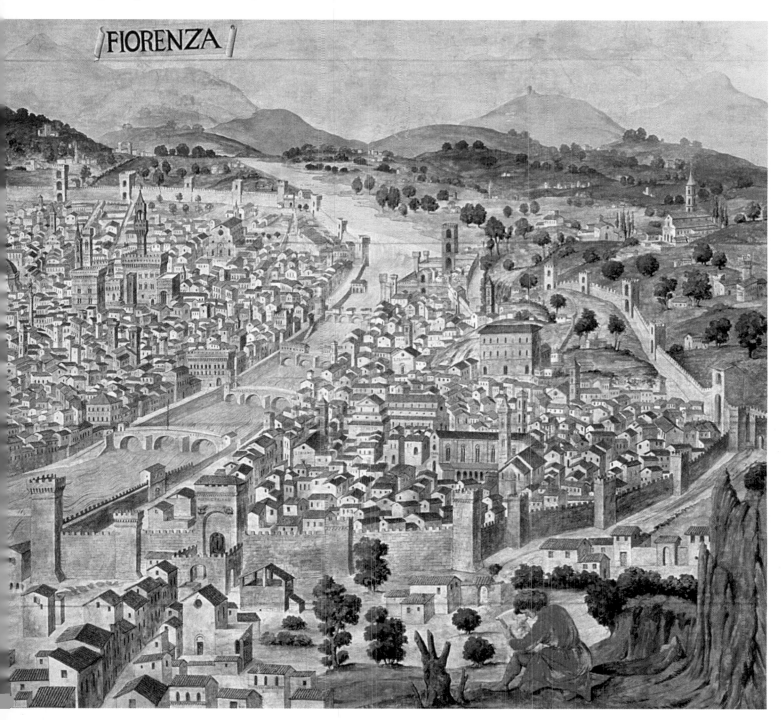

FIORENZA

Florence, the city of David, must equal or surpass in magnificence the cathedrals found in Siena, Pisa, Milan, or Venice; by 1436 the signature dome of Santa Maria del Fiore had been completed.

The dome, the design of artist-architect Filippo Brunelleschi, was a marvel of engineering. At the time, the cathedral boasted the largest dome ever built; even today, its dome is the largest masonry dome in the world. To construct a dome of such dimensions, Brunelleschi came up with an ingenious double-walled design that did not require centering—the technique of using a temporary support structure under a dome or arch and removing it only after the arch or dome is joined at the top. Brunelleschi built hoists and cranes that raised countless bricks—more than four million in all—some 350 feet in the air.

Thirty-five years after the completion of the dome, Andrea del Verrocchio's workshop was awarded the contract to place a gilded ball measuring eight feet in diameter on top of the dome. Verrocchio's 19-year-old apprentice, Leonardo, was intimately involved in this effort. On May 29, 1471, Leonardo climbed to the top of the dome, and, using hoists similar to Brunelleschi's original designs, assisted in putting the ball in place. His lifelong obsession with engineering may have begun with this project.

The Medici, key backers of the new dome and gilded ball, were the most powerful and important family in Renaissance Florence. By the early 1400s, Cosimo de' Medici, known as Cosimo the Elder, and his father, Giovanni di Bicci de' Medici, had founded a banking house, with branches in all the important city-states of northern Italy, which established the family's wealth and preeminence. After the accession of Cosimo the Elder to the office of Gran Maestro (head of state) in 1434, his direct descendants ruled Florence, with only a few short periods of exile following popular uprisings, until 1537. Cosimo's son Lorenzo came to power in 1469 at the age of 20, after the brief and troubled reign of his older brother Piero that had followed the death of Cosimo in 1464. Lorenzo became one of the most capable leaders of Renaissance Italy and figured prominently in the life of Leonardo during his time in Florence.

A World of Intrigue

Danger was never far from the lives of the Medici family. On April 26, 1478, Giuliano de' Medici, brother of Lorenzo, was stabbed to death by Bernardo di Bandino Baroncelli while attending mass in the Duomo. Lorenzo, sitting nearby, was attacked by two other assassins. Lorenzo was wounded in the neck, but he escaped with his life. The attacks were almost immediately revealed to be part of a plot by a rival merchant family, the Pazzi, who were backed by Pope Sixtus IV and other anti-Medici interests. Reprisals began immediately. The leaders and conspirators were hunted down mercilessly. On the first night after the attacks more than 20 men were hanging from the windows and loggias of public buildings in Florence. The wounded Lorenzo, appearing at a window of the Palazzo Medici, encouraged the bands of marauding vigilantes in the streets below. Verrocchio received a commission to do a wax figure of the wounded and bandaged Lorenzo, and the painter Botticelli did several portraits of the hanging traitors. These works do not survive.

Above: Giotto oversaw the construction of the base and the first of six levels of the Duomo's campanile, a 278-foot-tall bell tower, before he died in 1337. Visible in the background is Brunelleschi's majestic dome, completed a century later.

Opposite: The view of Florence from atop the 375-foot-high dome of the Duomo reveals a sea of terra-cotta-tiled roofs, many on buildings surviving from Renaissance times.

"Ghiberti's 'Jacob and Esau' on the famous Baptistery doors in Florence shows perspective used to achieve a spatial harmony that has almost a musical effect."

—KENNETH CLARK

CIVILIZATION, 1969

Above: Among the ten panels of Ghiberti's Doors of Paradise, panel 6 (right door, third one down) is a collage of Abel tilling the fields, the "Curse of Cain," and the deliberation of the judges.

Opposite: Visitors brave a light rain in order to admire Ghiberti's Doors of Paradise, which depict stories from the Old Testament.

FLORENCE'S DUOMO

European religious architecture of the Middle Ages, whether sponsored by church or state, usually represented a very lengthy commitment in time and consumed the lives of generations of workers. As early as the late fourth century A.D., a church already stood at the site now occupied by the Duomo in Florence. In the seventh and eighth centuries the Florentines replaced the ancient church with the Chiese di Santa Reparata and began the construction of a baptistery. By 1128, with the growing population of the region, the Santa Reparata had gained the status of a cathedral and once again needed to be enlarged. But it would be close to 170 years before construction began on the new Duomo di Santa Maria del Fiore (Cathedral of St. Mary of the Flowers)—destined to become the architectural centerpiece of the city. Prominent sculptor-architect Arnolfo di Cambio was put in charge not only of drawing up plans for a drastically expanded cathedral but also of reorganizing the growing city.

With the death of Arnolfo in 1302, all construction activities were suspended. When construction resumed on the Duomo in 1332, a bell tower was added to the plans. For that project the greatest contemporary painter, Giotto, was commissioned to design the tower and to personally oversee its construction. Indeed, Giotto supervised the work on the lower courses until he died in 1337. Two younger contemporaries—architect Francesco Talenti and sculptor-architect Andrea Pisano—finished the tower according to Giotto's design. The tower displays multiple courses of increasing length and is adorned with variegated marble slabs, and bas-reliefs carved by Andrea Pisano.

Even when, in 1394, the Milanese, the city's powerful neighbors to the north, laid siege to Florence—in an attempt to annex the city and make it a part of a unified northern Italy—construction activities continued on the cathedral complex. In 1401-02 the Cloth Merchants Guild had even invited artists to submit proposals for the gates of the north door of the Baptistery.

Nine years after the Milanese invasion began, their leader died of natural causes and his army lost the appetite for warfare. Their quest was abandoned. In celebration, the Florentines increased the pace of artistic and architectural activity already afoot, offering new commissions for work that would enhance and beautify the city.

In 1418 the city buzzed with excitement over the awarding of a commission to complete the dome of the cathedral. A century before the Milanese invasion, Arnolfo, the original architect, had left the design and construction of the dome for a future generation. The building now had a temporary wooden dome. The Wool Guild posted an extremely generous prize of 200 florins for the "design for the vaulting of the main dome of the Cathedral." The dome would have to be as expansive as the Pantheon in Rome, but set on top of the massive building. Artist-architects Lorenzo Ghiberti and Filippo Brunelleschi, finalists in the design competition for the baptistery doors, were again the finalists, and this time the commission went to Brunelleschi instead of Ghiberti. Inspired by the Pantheon, which he had studied while in Rome, he proposed a dome evocative of the narrow end of an eggshell, but with an octagonal cross section, to be built entirely without the use of centering or interior support while under construction. To carry out his plan Brunelleschi invented ingenious engines of architecture—hoists that could raise countless bricks, suspend them above the rising dome, and allow bricklayers to weave the dome one brick at a time. The dome was finally completed in 1436.

Above: The interior view of Brunelleschi's dome, illuminated by light from the octagonal lantern topping the dome. Depicted in concentric rings are scenes from the Last Judgment painted by Giorgio Vasari and Federico Zuccari.

Opposite: The Duomo and its dome, reflected in the well-polished hood of a modern automobile.

Benozzo Gozzoli, "Procession of the Magi," 1459-1462. In depicting the magi on their journey to Bethlehem, the artist used prominent Florentines as models, including Cosimo de' Medici, patriarch of the prominent banking family, and his son, the young Lorenzo de' Medici, pictured on the leading white horse.

One small sketch associated with the incident, however, does survive. More than a year later, in late 1479, the murderer of Giuliano, Bernardo di Bandino Baroncelli, who had escaped to Constantinople, was arrested by officers of the sultan and returned to Florence where he was tried, sentenced to death, and executed. Leonardo witnessed the execution of Bandino in the Piazza della Signoria on December 28, 1479, and in one of his *libricini* (small notebooks) made a sketch of the assassin hanging from a window of the Bargello. Then with characteristic acuity, he described the event, and even the clothes Bandino was wearing. In the manner of a modern photojournalist recording an event, he wrote: "black silk jacket; fox

hat black, the color of the jacket; cape of blue satin, fur lined; beard …." Then he added, "Strange! Why, he is still dressed as a Turk." The incident revealed Leonardo's disinterested curiosity as well as his powers of observation and emotional detachment—critical qualities that would serve him well in his anatomical studies.

Beginning with Lorenzo de' Medici's grandfather, Giovanni, who supported the artist Masaccio, and Lorenzo's father, Cosimo, who prized the work of Donatello, the Medici were generous patrons of art and architecture in Florence, commissioning many of the great works that we associate with Florence and the Renaissance today. Lorenzo ordered work by Verrocchio, Botticelli, Ghirlandaio, Leonardo, and Michelangelo. A great bibliophile, Lorenzo continued the task, begun by his father, of amassing a magnificent library; the Laurentian Library was one of the best of the age. Lorenzo also supported the work of many scholars and humanists, including that of his friends Marsilio Ficino, founder of the Florentine Academy, and Giovanni Pico della Mirandola, author of the famous statement of Renaissance sensibility, *On the Dignity of Man*. And as a statesman, Lorenzo used careful diplomacy to help establish a delicate balance of power among the Italian city-states that preserved the peace of northern Italy until shortly after his death in 1492.

After 1537, the branch of the family descended from Cosimo the Elder's younger brother, Lorenzo the Elder, came to power, producing the important Renaissance popes, Leo X, Clement VII, Pius IV, and Leo XI, through whom the family controlled both Florence and Rome. The Medici continued their patronage of art, architecture, and literature, endowing Florence with gems such as the incomparable Galleria degli Uffizi (Uffizi Gallery) to house their art collections.

The spirit of competition in the arts that flourished among the city-states of Renaissance Italy was mirrored in politics and economics. The city-states, along with the Papal States of central Italy, and outside powers, such as the Holy Roman emperor and the French monarchy, competed for political ascendancy on the Italian Peninsula. It was not a game for the fainthearted. No individual city or collection of cities was able to exert lasting control over the country. (Not until the 19th century was Italy unified as a nation-state.) Alliances were made, broken, and readjusted to ensure that the strongest and most aggressive powers were held in check by combinations of the others. It was a system concerned primarily with an unsentimental and rational pursuit of a balance of power—and it did not always work. Warfare or the imminent threat of warfare was the norm; treachery was rife. Friends quickly turned into enemies. Within individual cities, no leader was entirely safe, even from members of his own family. The poison vial and the dagger were common tools of politics. In this unforgiving political environment, Leonardo's Florentine friend Niccolò Machiavelli wrote *The Prince*, the statesman's handbook for survival. Artists may not have been in positions as hazardous as those of the politicos who were their patrons, but, to prosper, an artist had to keep a finger in the wind, especially as the 1400s drew to a close.

At the beginning of the 1500s, artists and sculptors were regarded as craftsmen, on a par with other artisans such as goldsmiths. Generally artists were not drawn from the higher classes of society; like artisans, they were educated for a trade. Artists were taught to read and write in the vernacular; given enough mathematics to keep accounts and carry on a simple business; and at 11 or 12 years of age, apprenticed to the workshop of a master, where they learned to draw, paint, and sculpt. Such was probably the early education of Leonardo, even as late as the 1450s and 1460s. In 1400, painters and sculptors would not have been welcome at the court of a duke or king. Most of their works would have been regarded as devotional objects or decoration. By the end of the century, however, many—including Leonardo—were courtiers, welcome in the company of the learned and powerful and expected to be conversant in public affairs. Moreover, their work was revered, and considered to be on a par with music and poetry.

Above: Leonardo, 1479. Pen-and-ink sketch on paper. Musée Bonnat, Bayonne, France. In 1479, Leonardo witnessed the execution of Bandino, the assassin of Giuliano de' Medici; he sketched the event in one of his notebooks.

Following pages: Originally designed by Vasari, the Galleria degli Uffizi in Florence is one of the oldest museums in the world and home to an unrivaled collection of Renaissance art. The long, narrow courtyard, evocative of a street, separates two symmetric wings of the Uffizi.

At court, an artist had to be politically nimble, as ready as any prince or diplomat in Italy to switch allegiances and steer a fresh course if interests so dictated—and no artist adapted to this environment better than Leonardo. If his relationship with one patron or city became strained, he quickly ingratiated himself with another. If a patron fell from power, he won the favor of another. For Leonardo, the change in the status of the artist led him not only to exercise his considerable verbal and diplomatic skills but also to stretch his intellect. Though a lifelong pacifist, he served as a military engineer designing futuristic weaponry. In the 1480s, while attached to the court of Ludovico Sforza, strongman of Milan, Leonardo served as musician, dramatist, and set designer for court theatricals. But despite embarrassing scandals and a reputation for not finishing commissions, Leonardo always landed on his feet, managing to maintain himself in surroundings that allowed him to pursue his passions and intellectual interests.

"In this labor ... I have undertaken ... the writing down the lives, the works, the manners, and the circumstances of all those who, finding the arts already dead, first revived them, then step by step nourished and adorned them, and finally brought them to the height of beauty and majesty whereon they stand at the present day."

—VASARI

Giorgio Vasari, "Self-portrait," ca 1567. Oil on canvas. Galleria degli Uffizi, Florence. The first and still regarded as the greatest of art historians, Vasari was a highly competent artist and architect, though he freely admitted never being in the same class as many of the immortal artists whose lives and works he portrayed in his monumental book, The Lives.

In 1546, Italian painter Giorgio Vasari—a beneficiary, even more than Leonardo, of the sea change in the status of art and the artist—attended a party in the palace of Cardinal Farnese in Rome. Discussing the staggering output of Italian, particularly Florentine, architecture, painting, and sculpture over the preceding 150 years, Vasari realized that the story must be told: The story of how medieval art, beginning with Giotto, was gradually transformed into the masterpieces of Leonardo, Raphael, and Michelangelo, and how the plastic arts were elevated from crafts to expressions of humanity's deepest aspirations. The result was Vasari's famous work, *The Lives of the Most Eminent Painters, Sculptors, and Architects,* published in 1550 and revised in 1568—and still regarded as art history's most important work.

The Status of the Artist

The elevation of the social and intellectual status of the artist in Italy began in the late 1300s in the city of Padua. Petrarch, the great poet, humanist, and educational reformer of the 1300s, had made his home in Padua. Often called the father of humanism, Petrarch died in 1374, but his legacy lived on, and Padua reached its intellectual zenith during the final years of the century. It was there, in the 1390s, that Cennino d'Andrea Cennini wrote *Il Libro dell'Arte (The Craftsman's Handbook)*—the first work in Italian detailing the craft of painting. Cennino was not a master of the written word, but his text was the crude beginning of an extended written discussion of painting that would span the next 150 years. Such writing became more consciously literary and technically sophisticated during the 1400s. By the early 1500s the artistic treatise was a recognized genre.

Though he wrote no treatise as such, Raphael's letters and other literary productions were of interest for his comments on his art and his views about artists. Even Leonardo's intermittent musings about the art of painting, scattered throughout his notebooks, were assembled after his death by his pupil and executor, Francesco Melzi, into *Il Libro di Pittura,* or *Treatise on Painting.* By then it was taken for granted that others would be keenly interested in the thoughts of a master. In little more than a century, the painter had evolved from trained craftsman to specially gifted artist.

Cennino's *Handbook* reads for the most part like a straightforward textbook for an apprentice artisan, which is what all painters were considered to be in the 1390s. Future artists arrived at a *bottega,* or workshop, of a master in their early teens after the brief education given to those destined for apprenticeship in a craft. Few came from the elite classes. Most were the sons of the artisan class. This remained true well into the 1400s. But Cennino's *Handbook,* though primarily dealing with questions of craft, had taken the first steps, in a few passages, toward asserting the importance of the imaginative life and creative genius of the artist.

The writers on art who followed Cennino, most of whom were themselves practicing artists, slowly chipped away at the crabbed, narrow conception of the artist, at the same time expanding the possibilities of art, especially painting, through the introduction of new technology and

GUTENBERG'S PRESS

Johannes Gutenberg developed alloys and molds necessary to produce precise, uniform, moveable type that, with an adapted winepress, made the mechanization of book production possible. By 1450 Gutenberg had an operating printing press. The first printed text may have been a poem printed on single sheets. There is speculation that around 1450 his press printed a Latin grammar. By 1452 he was probably at work on the famous 42-line Bible. The book appeared in 1454-55 in an edition of about 180 copies. It is still among the most beautiful examples of the craft of the printed book. It sold for 30 Italian florins, a sum representing at least three years' wages for an average worker. But a similar handmade Bible would have sold for a greater amount and taken over a year for scribes and illuminators to produce. With his first printing venture, Gutenberg demonstrated the economic viability of his invention.

innovative techniques. The workshops, where these writings were treated with the greatest respect, became centers of experimentation and exchange of ideas. The clients of the workshops were also changing. They were more numerous and more prosperous and—immersed in the new humanist outlook—they were demanding a different kind of art. The workshops were ready to oblige. When Leonardo arrived in the bottega of Andrea del Verrocchio in Florence in the 1460s, the transformation was well underway.

Early in the 1400s, artist-architect Filippo Brunelleschi, who would complete the dome of Santa Maria del Fiore a few decades later, wrote a short letter on the proper use of perspective in painting. The text has been lost, but the frequent mention of it by other writers and artists attests to its influence. The careful description of this technique and its relatively swift adoption by most working artists created one of the key stylistic markers that separate medieval from Renaissance painting. Piero della Francesca, a painter who apprenticed in the Florentine workshop of Domenico Veniziano before 1430, in addition to displaying the new humanist sensibility in his work, also incorporated geometric forms and nearly perfect use of the new technique of perspective. His treatise, *De Prospectiva Pingendi,* or *On Perspective for Painting,* became the first classic statement of the technique—and a text familiar to artists of his day. Piero knew Brunelleschi and the painter Masaccio, another early master of perspective. Piero was an avid mathematician who encouraged artists to learn geometry. His *Short Book on the Five Regular Solids* had a strong influence on the work of Fra Luca Pacioli, a mathematician with whom Leonardo would later collaborate as an illustrator.

But the seminal treatise for the new artists of the mid-century, the work that extended existing boundaries and claimed whole new territories for the painter, was Leon Battista Alberti's *De Pictura,* or *On Painting.* It was written originally in Latin, circa 1435, as a textbook for use at the Casa Giocosa, a school for the elite of the court in Mantua. The book discusses all the latest advancements in paint and painting, including perspective, and pleads for a broader and more generous education for the modern painter. Alberti makes a strong case for the high regard in which painting and the painter were held in classical times. (The humanists by now had succeeded in elevating the pronouncements and customs of the ancients to a high level of authority.) Both the content of the book, and the audience for which it was intended, attest to the great change in the status of the artist in just the few decades since the 1390s. Almost immediately it was translated into Italian. The Italian edition was dedicated to Brunelleschi, Donatello, Luca della Robbia, and Masaccio, the first generation of great artists of the Italian Renaissance.

Alberti's influence was felt throughout the rest of the Renaissance and well beyond Italy. He told artists to look to nature as their guide, to experiment with new methods and materials, to educate themselves as thoroughly as any prince or courtier. They listened—none more seriously than Leonardo, who owned a copy of the treatise and for whom Alberti was a guiding spirit. It was a time of great ferment in the workshops, which sometimes tried to keep their latest discoveries secret in order to prevent them from falling into the hands of competitors. This was rarely successful. New methods in one workshop usually found their way into others. Many important techniques and technologies emerged before the end of

Above: Albrecht Dürer, "Young Hare," 1502. Watercolor. The grandson of a successful printer and publisher and son of a gifted painter, Dürer traveled to Italy on two occasions to study advanced techniques, especially linear perspective. The "Young Hare" demonstrates implicit understanding of two-point perspective. His monogram appears at center bottom.

Opposite: Dürer, "The Festival of the Rosary," 1506. Oil on panel. Dürer painted this piece in Venice for German residents of the city.

the century, including oil paints and painting on canvas—and Leonardo had a hand in almost all of them.

The Artist's Secrets

During the early Renaissance, the bottega was an all-purpose workshop for the decorative arts—where metalwork, stage sets, and furniture were as likely to be commissioned as paintings, murals, and statuary. For the variety of its functions, the bottega was also an early laboratory for experimentation in qualitative and

"The first intention of the painter is to make a flat surface display a body as if modeled and separated from this plane."
—LEONARDO

quantitative chemistry. The results of these experiments were closely guarded industrial secrets of these early laboratories. Although interaction with other groups and diverse industries was rare, it did take place in later times, and with dramatic symbiosis. For example, the exchange of ideas between the bottegas of Venice and the workshops of the glassmaking industry flourishing on nearby Murano Island bore fruit for both groups. When Venetian artists, especially Titian and Giorgione, introduced hitherto unseen vibrant colors, Raphael hired away a highly skilled Venetian studio worker in an attempt to plumb the man's workplace secrets. And Raphael's paintings immediately started displaying the more brilliant colors seen in the Venetian School.

Grinding the ingredients to create the pigments for paint, preparing the surfaces on which the paint was to be applied, and producing the tools, including the paintbrushes from animal hair, were all routine responsibilities of apprentices. For a steady supply of eggs used in painting in tempera, workshops even had their own well-stocked chicken coops. Each bottega had its own set of formulas, its own set of sources for color, but around 1390, Cennino's book detailed many trade secrets and provided a launching pad for advances in the technology of the artist.

Although it was the master who designed and finished the preponderance of a painting, a collaborative effort was the usual order in the studio; certainly the senior apprentices were frequently assigned the task of painting parts of the master's works. Leonardo revealed his genius early, in his apprentice days in Verrocchio's studio; from the beginning, his work is quite unmistakable. Later when he opened his own studios, first in Florence, then in Milan, he had his own apprentices, and their work also is unmistakable.

The art of medieval times had been exclusively devoted to religious subjects, with symbolism far more significant than reality. Artistic renditions were flat, the subjects' faces stiff and devoid of emotion. When we view them now, they do not communicate to us; they are not psychologically charged. Also missing are artistic devices for conveying depth or linear perspective. The effect of the atmosphere, specifically the degradation of light with increasing distance, is not taken into account; nor is the subtle blurring of the curved edges of the subjects and objects. The underlying subtle hues of color in the skin, colors in shadows, and the hues of blue in snow and water are all absent.

In the 15th century, painters were applying progressively paler colors on receding objects to create a sense of distance. They introduced the techniques of chiaroscuro (meaning "light and dark"), using light and shade subtly along the edges of a form to give it definition, and sfumato (from the Latin root *fumus*, meaning "smoke"), blurring the edges of distant objects to blend them seamlessly into their surroundings. Early art historians attributed both techniques to Leonardo, but chiaroscuro had been introduced almost a generation before Leonardo, and sfumato most likely appeared first in the workshop of Andrea del Verrocchio, where Leonardo received his training. Where Leonardo did not invent these techniques, he perfected them, as a result of his unending studies of light and shade on curved surfaces. One sees the evolution of his few—but sublime—paintings from simply beautiful works, which may lack flawless perspective, to masterpieces: In his early "Annunciation" (ca 1472-75), the perspective is imperfect; in "The Last Supper" (1498), the composition is far more complex, and the perspective is perfect—a result of his preoccupation with mathematics.

By the time he painted the "Mona Lisa" (ca 1507), he had gone beyond one-point perspective to two-point perspective—where the lines of perspective recede to a separate pair of vanishing points on the horizon line. Leonardo's subjects sit more naturally, and they move more naturally than in others' works—an understanding born of his anatomical studies on cadavers. Finally, there is a new element, unsurpassed to this day—the psychological charge that he gives his subjects. The inscrutable smile of the "Mona Lisa" makes it the most famous portrait in the world; in "The Last Supper" the thunderclap of electricity captured in Christ's announcement of his betrayal by one of his disciples earns acclaim for the painting as the "keystone of European art" by art historian Kenneth Clark, and universal acclaim as the first painting of the High Renaissance. The best guess after all these years is that Leonardo possessed both a photographic memory and a preternatural vision, allowing him to freeze motion in the air. Above all, he possessed the keenest powers of observation ever demonstrated by a painter.

RENAISSANCE SENSIBILITIES

A new view of humanity had begun to emerge at the beginning of the 14th century, one that regarded each individual as uniquely promising in his or her relationship with other men and women, and with God and his universe. This new sensibility looked to the poets, dramatists, orators, artists, and philosophers of the ancient past for wisdom and inspiration. It looked to the future with aspiration, with a feeling that humankind was just beginning to realize its potential. The great Dutch humanist Erasmus caught the new mood when he wrote "people run into the churches as if they were theaters, just for sake of the sensuous charm of the ear." That same spirit was captured in the Cantoria, or Singing Gallery, of the Duomo in Florence, adorned with a set of square bas-reliefs with the theme "Joy of Children," carved by Luca della Robbià between 1431 and 1438. "Blast of Trumpets" is shown below.

Giotto's Legacy

In his *Lives,* Vasari identifies Giotto di Bondone as the first truly great artist deserving inclusion in his compendium. Realism, or depiction of subjects as they naturally appear, had, according to Vasari, been ignored in the works of artists for the preceding 200 years; Giotto, with his surpassing painterly prowess, restored this quality and paved the way for generations of great Florentine artists to follow. The series of 37 murals painted between 1303 and 1310 in the Scrovegni

Giotto di Bondone, "Flight into Egypt," ca 1305. Fresco. Scrovegni Chapel, Padua. One of several frescoes in the chapel depicting the life of the Virgin, this fresco illustrates the episode when, following King Herod's decision to massacre newly born male children as potential future rivals to his throne, angels appear to Joseph, advising him to flee.

Chapel, Padua, is regarded as Giotto's magnum opus. Other masterworks include the "Legend of St. Francis's Confirmation of the Rule by Innocentius III" in the Basilica of Assisi, though the authorship is disputed; and the 'Ognissanti Madonna" (ca 1314), named for the Church of the Ognissanti in Florence. The quality that distinguishes these paintings from the works of other painters of the period is precisely the realism with which Giotto was able to inform his subjects. According to Leonardo's own explanation, "[Giotto], who being born in the mountains in a solitude inhabited only by goats and such beasts, and being guided by nature to his art, began by drawing on the rocks the movements of the goats of which he was the keeper." Indeed, Leonardo could have been speaking of himself.

Linear Perspective

One of the most powerful techniques for achieving realism in painting is that of effectively projecting a three-dimensional space of reality onto a two-dimensional surface, and conversely of making a two-dimensional image appear to have depth. The underlying idea in linear perspective is that in observing a painting, the viewer is meant to "see" the distance in precisely the same manner that one would in looking through a window pane.

The horizon line represents the eye level of the artist. This is also the eye level of the viewer of the work of art. Peering into a room or a closed courtyard, where the facing walls are parallel to each other, all of the parallel lines—those representing the horizontal edges of pictures or tapestries hanging on the wall; the sills of windows and doors; the seams of floor tiles; and the lines created by conjunction of walls and ceilings, and walls and floors—will appear, when extended (or extrapolated), to converge at a single point on the horizon line. This is the compositional technique known as one-point perspective. Linear perspective is a technique with which artists in classical times may have experimented, but surviving evidence in the murals of Pompeii suggests that it was imperfectly understood.

According to historians of art, the proper understanding of one-point perspective is a Eureka! moment that came to Filippo Brunelleschi. The artist-architect journeyed to Rome in 1405 for a visit that turned into a 13-year stay. Lessons from Rome's Pantheon, a few decades later, inspired Brunelleschi's designs for the Duomo. And quite likely, his visits to the ancient Roman Forum—where he saw the remnants of classical buildings arranged in neat arrays, with varied shapes and sizes and rows of columns—gave him a sense of how to project a three-dimensional space onto a two-dimensional surface. On his return to Florence, Brunelleschi painted an inner wall of the Duomo's baptistery with correct one-point perspective. Soon,

Above: Andrea Mantegna, walls and ceiling of Camera degli Sposi, 1465-1474. Castello San Giorgio, Mantua. Mantegna, a master of linear perspective, painted the walls and ceiling of this small interior room evocative of an open-air pavilion. On the ceiling, a painted dome opens onto a sky, with viewers—men, women, and cherubs—appearing to look into the room.

Opposite: Mantegna, "The Dead Christ," 1490. Pinacoteca di Brera, Milan

Masaccio and Ucello began to invoke the technique in their works. Masaccio's "Tribute Money" is an early work embodying superb linear perspective.

It took Ghiberti 27 years (1425-1452) to complete the bronze doors on the east side of the baptistery, next to the Duomo. The ten square panels in bas-relief, sculpted designs that stand out from the surface, took so long in their production that the process paralleled the experimentation going on with Brunelleschi's newly demonstrated idea of linear perspective. Accordingly, the panels reflect an evolution in artistic style. It is with the fifth and sixth squares cast in the late 1430s that the application of the theory of linear perspective

becomes most pronounced, where there is an appearance of greater depth in the image than in the earlier panels.

Leonardo made a rough sketch of a perspectograph, a device used to achieve effective one-point perspective, and in 1506 Albrecht Dürer (1471-1528), Germany's greatest painter, known as the "northern Leonardo," journeyed to Italy to learn

RAPHAEL'S "SCHOOL OF ATHENS"

This group portrait of ancient philosophers, whose lifetimes span some 1,500 years, is emblematic of the spirit of humanism fueling the Renaissance. The quartet standing at bottom right includes Greek astronomer Ptolemy, philosopher and poet Zoroaster, and a self-portrait of Raphael, peering out at the viewer. Including a likeness of one's self in a painting was a common artistic device of the time. Just to the left of the quartet and hunched with a compass in hand is Greek mathematician Euclid, demonstrating a geometric proof to his students. The model for Euclid is thought to be Bramante, the architect of St. Peter's Basilica. The trio on the top step in the upper left comprises military genius Alexander the Great, Greek historian Xenophon, and Greek philosopher Socrates.

The two central figures on the top step are Greek philosophers Plato and Aristotle. Plato (left) holds a copy of his dialogue, the *Timaeus,* and Aristotle a copy of his treatise, the *Ethics.* Although the model for Artistotle is unidentified, Plato bears an uncanny resemblance to Leonardo. This is ironic since Leonardo was more a pragmatic Aristotelian than an idealistic Platonist. Finally, near the front center, the dark, brooding figure with an elbow on a chest is meant to be ancient Greek philosopher Heraclitus—the model, Michelangelo. In Raphael's early cartoon (full-size drawing) of the painting, Michelangelo is missing. However, after Michelangelo expressed his deep admiration for the mural, Raphael added the figure of Heraclitus in the likeness of Michelangelo. Raphael may

have decided to depict the pre-Socratic philosopher, known as an exceedingly difficult and irascible man, since his own great contemporary, Michelangelo, was known for similar qualities.

Raphael sets his assemblage in an idealistic space, cobbled together from the best of Roman and Renaissance architecture, and possibly inspired by Bramante's design for the new St. Peter's Basilica. The composition displays flawless one-point perspective. The lines delineating the floor tiles, ceiling cornices, and the radial pattern under the first arch all converge at a point, hip-high between the figures of Plato and Aristotle.

Raphael, "School of Athens," 1509-1510. Stanze di Raffaello, Apostolic Palace, Vatican Museums

the technique, and even produced woodcuts showing how a three-dimensional object can be mapped onto a horizontal two-dimensional board.

By the end of the 15th century, Andrea Mantegna (1431-1506) was experimenting with perspective and dramatic foreshortening when he produced the powerful painting "The Dead Christ" (1490). In 1509-1510, when Raphael (1483-1520) created his masterpiece, the "School of Athens" mural in the Vatican, he demonstrated impeccable one-point perspective. By this time, the realism achieved is far superior to that of Giotto's works of two centuries earlier.

For Renaissance artists, the discovery of linear perspective was so revolutionary and so exciting that they extended its application to other areas of artistic creativity. In 1481-82 Leonardo took a position in Milan, and there befriended the preeminent artist-architect Donato Bramante, who had recently been hired to design the Church of San Satiro. Bramante had seen in this project a compelling challenge. Because of external space limitations, the building could only support the shape of a T. But the prevailing style for churches in Lombardy called for a Greek cross floor plan with four equal arms. Accordingly, Bramante created an optical illusion, the appearance of an arm behind the altar—in the manner that a painter would have depicted it. When entering the church, one sees the arm behind the altar as an arm of normal length. Only in approaching the altar, however, does one encounter Bramante's joke. The apse is only a few feet deep.

The Technology of the Painter

In Leonardo's life, which spanned the Middle Renaissance and ushered in the High Renaissance, artists painted on wooden panels. These had to be prepared methodically. The panels were dried; then they were sanded and covered with a layer of liquid gesso—a coating made by mixing powdered gypsum with glue made from rabbit or other animal skins. The wood panels would then be rubbed vigorously with polished hematite, known for its hardness, until the desired smoothness had been achieved. Leonardo, the consummate scientist, described one of his own favorite gessoes in a notebook, a formula markedly different from the more traditional approach for preparing a panel for egg tempera: "Coat [the panel] with mastic and white turpentine of the second distillation … then give it two or three coats of aqua vitae in which you have dissolved arsenic or some other corrosive sublimate … apply boiled linseed oil so that it will penetrate every part, and before it cools rub it well with a cloth to dry it. Over this apply liquid white varnish with a stick, then wash with urine."

As for the paints, Leonardo, like other Renaissance artists of his time, first used egg yolk as the binding agent for the powdered colored pigment to create the medium known as egg tempera. He used minerals and other materials in concocting different colors and blended them according to specific recipes he had mastered in Verrocchio's studio, but which he continually improved throughout his life. Egg tempera had a shortcoming in that it could not be stored. Accordingly, each color was mixed as it was needed, and the artist had to work rapidly since the medium dried quickly. Mixing too little paint created problems: When the initial batch was depleted, mixing additional paint to achieve a match with the earlier was virtually impossible. And mixing too much paint was a waste of expensive materials.

"Perspective, with respect to painting, is divided into three parts … the first is the diminution in the size of bodies at various distances; the second part is that which deals with the diminishing in color of these bodies; the third is the diminution in distinctness of the shapes and boundaries which the bodies exhibit at various distances."

—LEONARDO

The masters built up the paintings not with mixed colors but in a prescribed order—thinned black ink to create a gray underpinning, then *terre vert* (a neutral green), followed by burnt sienna, then accent colors of high value, such as carmine red and ultramarine blue (worth more than gold). After linseed oil was introduced in northern Europe as a binding agent for the pigment, its use revolutionized painting. Venetian artists were the first in Italy to use linseed oil as a binding agent; they also introduced canvas as an alternative to the wood board. Unlike egg tempera, which is opaque, linseed oil paint is translucent, because the binding agent is clear,

like glass. The pigment is suspended in the oil so that reflection and refraction both take place within the oil. By building up layers of color to show light, artists created the appearance of depth more convincingly. Moreover, since oil paint dries much more slowly than egg tempera, the artist could work on several areas of a painting at one time. The powerful technique of sfumato would not have been possible without the use of linseed oil as the binding agent for the colored pigment. Leonardo mastered the technique in painting in oil as no one had before him.

In his treatise on painting, Leonardo explains how to achieve the blending of light and shadows seamlessly, and mentions this in connection with the process of sfumato. Only recently did a high-tech analysis finally confirm his explanation. Dr. Cecilia Frosinini, art historian at the Opificio delle Pietro Dure restoration laboratory in Florence, announced in 2007 that technology had made it possible

"Leonardo succeeded to perfection in expressing the doubts and anxiety experienced by the Apostles ... in the faces of all appear love, terror, anger, or grief and bewilderment...."

—VASARI

Leonardo, "The Last Supper," 1498. Santa Maria delle Grazie, Milan

to probe layer by layer a painting's surface; like the archaeologist, the art historian uncovers the "strata" of the painting—the thickness, the order, and indeed how the colors were applied on the panel.

The work examined was an oil-on-wood painting, "Madonna of the Yarn-winder" (1501), in the one version known to be by Leonardo's hand. The nondestructive bombardment of the surface of the painting by an ion, or charged atom beam, revealed that Leonardo, having forsaken the use of a palette on which to mix his paints or colors, applied extremely thin layers of color, one on top of the other, to create a rich texture and a virtual three-dimensional effect, in a technique known as velature. The advance preparation of the colors was essential for him, as was their dilution on the panel. And although he usually used an appropriate brush for the purpose, he frequently applied the thin layer of colors using his thumb, and sometimes other fingers. This is especially evident in his later works, including the "Mona Lisa." It seems befitting—entirely in the spirit of Leonardo—that it took the application of 21st-century technology to achieve scientific confirmation of what had long been suspected: that he had ignored the palette and used his fingers to paint.

The Enigma of Leonardo

Leonardo arrived on Italy's art scene as an apprentice in a Florentine workshop in the 1460s. Yet, from his earliest days as an apprentice, his art incorporated all the innovations and trends of the time. He had mastered perspective and thought that painters should be adept at mathematics and geometry. He exhibited an affinity for the new interest in landscape painting, using landscape deliberately and carefully as background in his two most famous portraits—the "Mona Lisa" and "The Last Supper." He was without peer in the use of portraiture to interpret and reveal individual character. He used oil paint, a new technology imported from French and Flemish painters, to replace tempera, a powdered pigment mixed with water and egg yolk that dried quickly and did not readily allow reworking of the surface. And he believed that artists should learn from nature and that painting and science were one.

Leonardo's genius as a painter is unquestioned. But the creator of the two most famous paintings in the history of art was in fact only a part-time painter. One might claim liberally that he touched around 17 paintings; with somewhat greater certainty, that he completed about a dozen; with absolute certainty, only six. In contrast, van Gogh painted some 900 works in less than ten years. Rembrandt produced a hundred self-portraits alone, 57 in oil.

When Leonardo was not painting, he was a consummate scientist and engineer. But it is ultimately the cross-fertilization of all his disparate interests that led to his miraculous creations, whether in the arts or the sciences. Indeed, he was an artist doing science—he painted the finest anatomical drawings ever created—and a scientist doing art. In his paintings "Virgin of the Rocks" and the "Mona Lisa," there are geological musings that attract the interest of modern geologists—stalactites and stalagmites, perhaps even the suggestion of subduction and upsurge of rocks; there is the degradation of light with distance, best understood by modern physicists and meteorologists; and there are the results of detailed anatomical studies, of great interest to the modern anatomist.

Leonardo the Bibliophile

Leonardo, a self-taught man, assembled in his time a respectable private library. Around 1492 he made a list of his holdings in his notebooks. This list attests to Leonardo's intellectual range. Thirty-seven books are mentioned, including standard religious texts, such as the Bible, philosophical works, ancient and modern, as well as science and mathematical tomes. But about half the list is literary in nature—grammars, works on rhetoric, classical authors such as Livy and Ovid in translation, and Italian poets such as Petrarch. In the fall of 1504, Leonardo records another catalogue of his books. The list has grown to 116 volumes—a very respectable number for the time—including some manuscripts on vellum, which would have been expensive and testify to his seriousness as a book collector. The new list includes Alberti's treatise on architecture, *On the Art of Building,* and the *Summa de Arithmetica,* a synthesis of the mathematical knowledge of the time written by mathematician Luca Pacioli, Leonardo's close friend. And Leonardo has expanded his collection of popular literature. There are chivalric romances, satires, dramas, and even erotic and bawdy poems for leisurely consumption.

He negotiated his way from one commission to the next, from one court appointment to another, as skillfully as the most accomplished diplomats of his time, often with surprising brazenness. He signed onto deals that were decidedly not to his advantage, which often led to trouble, especially when combined with his tendency to procrastinate and renege on contractual obligations. He fell back on his ability to coax and cajole. Like other diplomats, he made promises that he did not keep—and perhaps never intended to keep. Yet somehow he always managed to conduct himself in a way that allowed the widest latitude for the exercise of his talents and intellect.

Although in some important ways he was a man out of and beyond his own time and place, the realities—political, economic, intellectual, and artistic—of Renaissance Italy had a profound effect on the shape of his work as artist, scientist, and inventor. He embodied the soaring, heroic spirit of his time—a time remarkably congenial to an intellect like Leonardo's. A palpable excitement and optimism filled the air; assumptions were being questioned, rules broken. And Leonardo liked nothing more than challenging accepted notions and assumptions, his restless mind reaching toward the edge of the possible, and sometimes beyond. Even when he was wrong—about the possibilities of human-powered flight, for instance—he was asking the most essential and leading questions, pushing the boundaries of inquiry and possibility.

While Leonardo's existing manuscripts and notebooks, even in their unintentionally abridged state, constitute an extensive intellectual biography of his genius, much less is known about the more conventional details of his life. The most famous source is still Giorgio Vasari, author of the *Lives.* There were a few other unpublished earlier sources, familiar to Vasari and used by him; the most important was an anonymous manuscript known as the account of the Anonimo Gaddiano—"Gaddiano" indicating the early ownership of this work by the Gaddi family.

Vasari was eight years old when Leonardo died; in the *Lives,* he devoted only about 5,000 words to Leonardo, compared to 40,000 on Michelangelo, who was still alive when Vasari was writing. Vasari is often unreliable: He is not careful about dates, his judgments often show a Florentine favoritism and he is partial to the distorting use of narrative conventions that call into question the truth of some of his anecdotes. Yet his outline of Leonardo's life is the best we have.

Vasari was in his late 50s when he penned the second edition of the *Lives.* Already he had seen Leonardo's reputation in ascendance. Despite the flagrant hagiography, his remarks often appear prophetic, for Leonardo has become a figure of mythical stature—unrivaled artist, architect, musician, mathematician, engineer, anatomist, geologist, futurist. Unavoidable, however, is an element of mystery: Leonardo was a bundle of contradictions, all of which enhance and underscore his mystique.

Two columns of titles in Leonardo's hand, in his characteristic mirror script, list the holdings of his private library.

LEONARDO'S SKETCHBOOK

THE PAINTER OF THE SUBLIME ANGEL IN VERROCCHIO'S "Baptism of Christ" also displayed a penchant for depicting the grotesque and the goulish. Leonardo was obsessed with juxtaposing what he called *brutteza* (ugliness) as a counterpoint to *bellezza* (beauty), frequently placing the contrast in the context of age. Notebooks, always handy, would be pulled out immediately to record faces in the crowd—the common, the rare, the homely, the beautiful—anything worth looking at was worth sketching. They reveal the unfathomable levels of Leonardo's curiosity.

"A Man Tricked by Gypsies," ca 1493. Royal Collection, Windsor Castle, U.K.

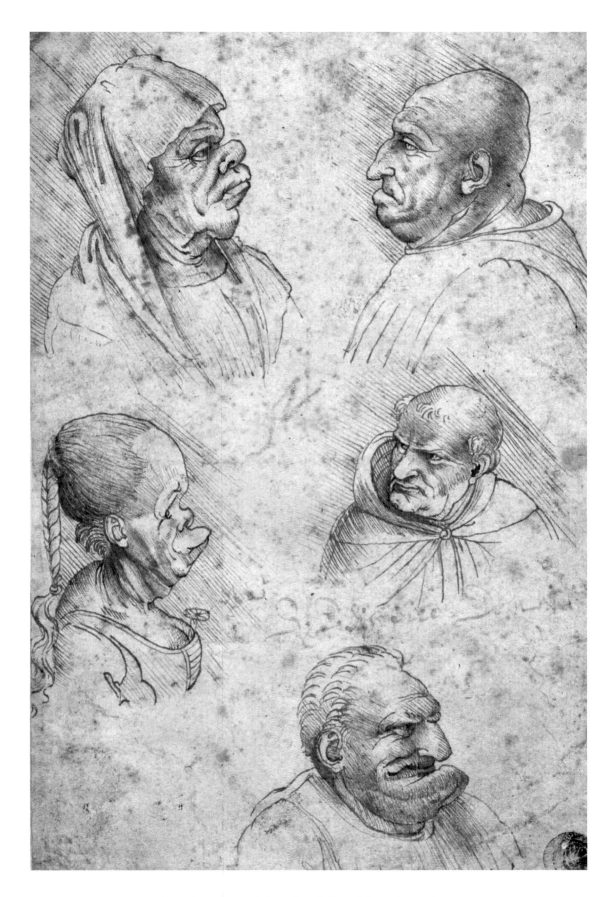

"Five Grotesque Heads," ca 1495. Galleria dell'Accademia, Venice

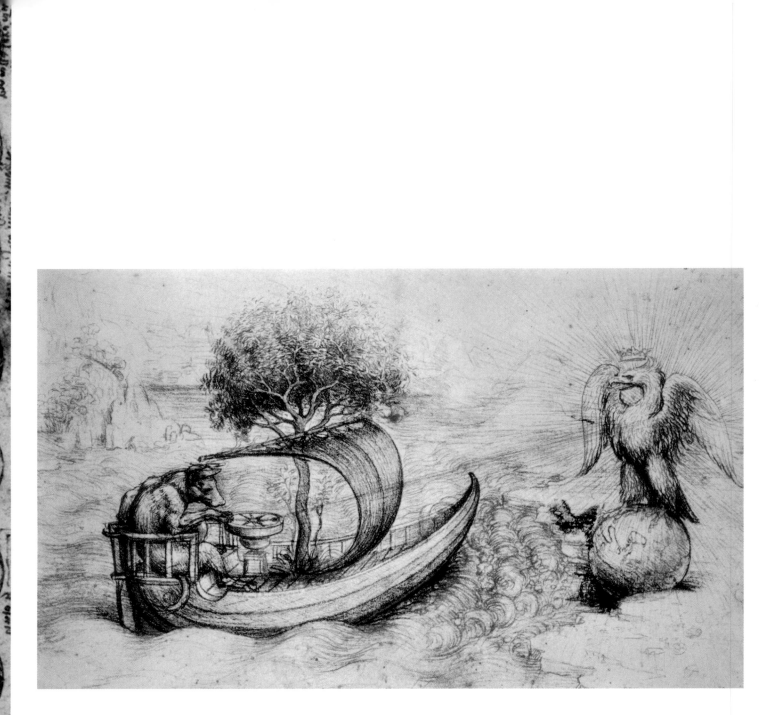

"A Political Allegory," ca 1495. Royal Collection, Windsor Castle, U.K.

"Neptune," ca 1504. Royal Collection, Windsor Castle, U.K.

"A Grotesque Caricature of an Old Woman," ca 1495. Red chalk. Royal Collection, Windsor Castle, U.K.

"Profile of a Woman," ca 1472-75 Gabinetto dei Disegni e delle Stampe degli Uffizi, Florence

Chapter Two

"There was born to me a grandson, the son of Ser Piero my son, on the 15th day of April, a Saturday, at the 3rd hour of the night. He bears the name Lionardo."

—ANTONIO DA VINCI

EARLY LIFE AND
FLORENCE

1452 Leonardo, the illegitimate son of notary Ser Piero da Vinci and Caterina, a farmer's daughter, is born April 15.

1453 Constantinople becomes Istanbul, capital of the Ottoman Empire.

ca 1465 Leonardo moves to his father's house in Florence and soon thereafter begins an apprenticeship with Andrea del Verrocchio.

1469 Lorenzo de' Medici becomes the ruler of Florence.

1471 Verrocchio's studio builds and mounts eight-foot gilded ball on Duomo in Florence; 19-year-old Leonardo is closely involved.

ca 1472-75 Leonardo begins work on the "Annunciation" and the "Madonna with the Carnation."

1473-75 Leonardo paints the "Portrait of Ginevra de' Benci."

1476 Leonardo establishes his own studio. He is denounced in scandalous Salterelli incident.

1478 Leonardo gets first recorded commission for an altarpiece.

1481 Leonardo undertakes "Adoration of the Magi" altarpiece and "St Jerome".

ca 1481 Leonardo departs Florence for Milan.

Little is known about the early years of Leonardo's life. The few reliable facts are often tantalizingly suggestive as to his later development and the evolution of his interests. There is always the temptation to make too much out of too little, and some analyses have pushed far afield. Conjecture of this sort about the boyhood of Leonardo is always interesting, often plausible—but, founded on such thin evidence, rarely conclusive. Far more can be said with conviction about Leonardo the man, a great deal based on his own words. Yet the frustrations of trying to penetrate his obscure boyhood never completely lift. At the beginning of his life, Leonardo evades us.

Love Child

What we do know of Leonardo comes from four contemporary biographers and maddeningly few documents, such as tax returns, deeds, parish records, and diaries. The most important source is the short biography of Giorgio Vasari in his *Lives of the Architects, Painters, and Sculptors*, first published in 1550. Vasari probably began research for the *Lives* in the mid-1540s, when Leonardo had already been dead for 25 years. Vasari would have relied on interviews with those who remembered Leonardo and on the few available writings about Leonardo that then existed; those included a short biographical sketch by a Florentine merchant, Antonio Billi, about whom little is known, and a short account by an anonymous author, usually referred to as Anonimo Gaddiano. A short account in Latin by Lombard physician Paolo Giovio also exists, written in the 1520s. Giovio may have known Leonardo personally, and he was present at the dinner-party discussion about artists and the art of biography in the apartments of Cardinal Farnese in Rome that inspired Vasari to take up the project of the *Lives*. Additionally, there is Giovanni Paolo Lomazzo, a painter who went blind in 1571 and thereafter turned to writing treatises on painting and commentary on important artists. He knew Leonardo's disciple

Opposite: The Castle of Vinci, as seen from the bell tower of the Church of Santa Croce
Previous pages: Ruins of the Etruscan city of Volterra and the verdant hills of Tuscany

and executor, Francesco Melzi, and was devoted to the memory of the great man. Through Melzi, Lomazzo had access to Leonardo's surviving writings and notebooks—and offers important insights.

Only one of these early biographers knew Leonardo, and then only late in his life. Probably only Lomazzo had access to Leonardo's writings. But their brief accounts, the surviving notebooks, and bits of documentary evidence that have come to light since their time are building blocks of the Leonardo biography.

One of these bits of documentary evidence was discovered in the 1930s. In a notebook that had been in the da Vinci family for several generations, Leonardo's

grandfather Antonio recorded the birth of his first grandson: "1452. There was born to me a grandson, the son of Ser Piero my son, on the 15th day of April, a Saturday, at the 3rd hour of the night. He bears the name Lionardo." Since the hour of the night referred to the time passed since sundown, the birth probably took place between 10 and 11 p.m. Grandfather Antonio also records the child's baptism and the names of the ten godparents who were present, a very generous number.

Leonardo's birth was illegitimate. His father was Ser Piero da Vinci, a professional notary, an occupation of some importance in the Italy of that time. Notaries drew up contracts, deeds, wills, and other legal documents. They were highly respected members of the community, hence the honorific "Ser" before Piero's name. He came from a long line of notaries: His grandfather and great-grandfather, though not his own father Antonio, had belonged to the profession. Leonardo's mother was a woman by the name of Caterina. Almost nothing is known about her. Vasari makes no reference to her. She did not marry Leonardo's father, and therefore it has long been thought that she must have been a peasant girl of lower social standing than Ser Piero, that this social difference accounts for the probability of her pregnancy being the result of a moment of lust, or the outcome of a passion that could not end in marriage. Though this sounds reasonable, it is only speculation. We simply do not know. The account of the Anonimo Gaddiano states that Leonardo was born of good blood through his mother. It is unlikely that such a claim would have been made on behalf of a peasant girl. But the author was writing in the 1540s, long after Leonardo's birth, and does not give a source for his assertion. Whatever may have been the basis of Ser Piero's relationship to Caterina, eight months after Leonardo was born, Ser Piero married a woman named Albiera, the daughter of a prominent Florentine notary. Perhaps he could not marry Leonardo's mother because this marriage had been previously arranged, something that would have been common at the time for someone of his social standing. It was certainly a far more advantageous match for him than Caterina. Ser Piero would now spend more time away from home as Albiera's family connections probably brought him profitable business in Florence.

Leonardo was most likely born in the family's house next to the parish church in the town of Vinci. A story arose in Victorian England that his mother was shuffled off to the countryside during her "confinement" to give birth to the shameful "love child" in the relative obscurity of the family's country house in the small village of Anchiano. But this story has no real support. It misses on at least three counts: Anchiano is never mentioned as a possible place of his birth in any account of his life prior to the 1840s. The house and property in question were not owned by Leonardo's family until Leonardo was older than 30 and long gone from Vinci, and—most telling—it projects back onto another time the sensibilities of a different, present time. Illegitimacy in the Italy of the 1400s did not carry the same stigma as it did in Victorian England. Many illegitimate sons of important families in 15th-century Italy rose to prominent positions in church and state. Leonardo's birth was certainly a welcome and celebrated family event, as can be seen from the tone of grandfather Antonio's diary entry and the ten godparents at his baptism. And Leonardo was to become the adored nephew of his childless Uncle Francesco. When dealing with Leonardo we can avoid many mistakes such as this by remembering that he is not of our own time, though he sometimes seems more a man of the 21st century rather than of the 15th. Keeping him in context not only explains those things about him common to most men and artists of the late 1400s in Italy but also sharpens our understanding of those wondrous occasions when he levitates from the pages of his notebooks and floats several centuries ahead of his own time.

"He was a beautiful man, well-proportioned, graceful, and of handsome aspect. He wore a pink, knee-length robe, though at the time long robes were worn: he had beautiful hair, which came down to mid-breast, curled and well-combed."

—ANONIMO GADDIANO

Following pages: Wild red poppies grow in a wheat field in San Quirico d'Orcia in Leonardo's native Tuscany.

LEONARDO'S APPEARANCE

The painter who elevated the likeness of a not-too-extraordinary woman, the wife of a Florentine merchant, to iconic stature in his painting of the "Mona Lisa," the scientist who poured out his observations in reams of notes, is ironically shrouded in mystery regarding his own likeness, and about his personal thoughts and his personal demons. He shines a spotlight on a universe of subjects, but not on himself. This pervasive enigma draws us to him.

Vasari, a child when Leonardo died, knew the artist only as a titanic, almost mythical figure, whose reputation had grown during the three decades that had passed between his death and the publication of the *Lives*. He wrote: "In Leonardo da Vinci, [is] a beauty of body never sufficiently extolled...." Vasari also offered a woodcut likeness of the artist, somewhat crude—but the best this medium allows—based perhaps on a few likenesses that may still have been around but have not survived into our time. There is the long beard, and the incisive eyes. But the woodcut does not entirely comport with Vasari's unstinting praise of Leonardo's looks, bodily grace, and powerful physical presence that "commanded everyone's affection." And the early Leonardo biographer known as the Anonimo Gaddiano (ca 1530) described Leonardo as "a beautiful man, well-proportioned, graceful, and of handsome aspect. He wore a pink knee-length robe ... [and] had beautiful hair, which came down to mid-breast, curled and well-combed."

As a young man Leonardo had a reputation as a dandy, always fastidious about his dress and appearance, with rings on his hands and fine leather boots on his feet, his person scented with rosewater and lavender. According to Paolo Giovio, another of the early biographers, his manners were equally refined. He described Leonardo as "by nature very courteous, cultivated, and generous."

One of the earliest possible likenesses of Leonardo is Verrocchio's bronze statue of "David" (page 90). When viewed in profile, there is a tantalizing resemblance to the later portrait of Leonardo in profile (page 17) done by Francesco Melzi, Leonardo's young assistant. The similarity has inspired some modern scholars to speculate that the young Leonardo, then Verrocchio's apprentice, may have served as a model for the statue.

Many art historians also speculate that the shepherd boy in Leonardo's unfinished "Adoration of the Magi" (page 85) is a self-likeness. It was common for painters in this period to make cameo appearances in their own works. Other examples include Raphael's appearance in his "School of Athens" and Michelangelo's in the "Last Judgment."

The well-muscled character in Leonardo's famous Vitruvian man (page 129) is often proposed as a self-likeness of Leonardo in early middle age. There is no hard evidence to support this, but Vasari tells us that Leonardo was a not only beautiful, but also a powerfully built man, strong enough "that with his right hand he could bend the iron ring of a doorbell, or a horseshoe, as if they were lead."

It is, of course, the bearded man of his later years that most people call to mind when they think of the physical appearance of Leonardo. Raphael—in the "School of Athens" (ca 1509-1510)—presented a group portrait of 30 great philosophers whose lives spanned approximately 1,500 years. Leonardo, then in his late 50s, served as Raphael's model for Plato. Born in 1483, 31 years after Leonardo, the young Raphael gave the older artist an honored place in his masterpiece.

Two images of an elderly man with impeccable features appear in the opening pages of this book. The man depicted in the famous red-chalk "self-portrait," discovered in Turin just 200 years ago, is thought by a majority of scholars to be Leonardo. The quality of the portrait and the left-handed shading make the drawing an authentic Leonardo creation, but is it an actual self-portrait? Art historian David Alan Brown claims the drawing is from the wrong period and is not Leonardo, although he does agree that the portrait in profile produced by Melzi is almost certainly of Leonardo in his later years. But the red-chalk drawing has become iconic. If it is not Leonardo, it has become the image we most associate with Leonardo, the essential expression of how the maestro looks to us.

Opposite: Detail of Plato (left), modeled on Leonardo, and Aristotle from Raphael's "School of Athens"

Above: Detail of Leonardo's unfinished "Adoration of the Magi," with the shepherd boy at lower right

A Country Boy

We do not know exactly where or with whom Leonardo spent his first five years—in the house of his father's family or in his mother's home. After his birth in 1452, nothing is heard of him until 1457 when he appears in grandfather Antonio's *catasto*, or tax return. Antonio lists him as a dependent for which he would have been entitled to a tax exclusion of 200 florins. It is not certain whether Leonardo was actually living in Antonio's household, or whether he maintained regular contact with his mother, Caterina, who, a year after Leonardo's birth, had married a man who called himself Accatabriga, a producer of lime, used in agriculture and in making mortar and pottery. According to their tax records, they lived in nearby Campo Zeppi, where they were smallholders of more modest circumstances than the family of Ser Piero da Vinci. Even if Leonardo never lived with Caterina, Campo Zeppi was close enough to Vinci that frequent visits by her son would not have caused anyone great strain.

No matter exactly where or with whom he spent his early days, one thing is certain—Leonardo was a country boy. He grew up surrounded by open fields, woods, looming hills and cliffs, birds, insects, and animals, wild and domesticated. It is also likely that with Leonardo's father frequently absent on business, Ser Piero's brother, Francesco, who was 15 when Leonardo was born, assumed much of the emotional responsibility for the boy. Francesco was not a notary like his brother Piero, nor did he follow any other profession. Like their father Antonio, Francesco lived the life of a country gentleman, depending for income on the property holdings of the family in Vinci and the surrounding countryside. This gave him the leisure necessary to invest his attention in the young Leonardo. Their mutual regard was lifelong and when Francesco died, he left his entire estate to Leonardo. There has been much speculation about his relationship with his father, mother, stepmother Albiera, and stepfather Accatabriga, though not much hard information. There is little doubt, however, that during these formative years Leonardo formed a special bond with Francesco. They romped together across the Tuscan countryside, and their experiences led to Leonardo's lifelong love of nature and curiosity about the world and how it works.

At this tender age Leonardo already revealed an interest in dissecting dead animals, developing skills of observation and techniques for systematic experimentation. At times his imagination also took a ghoulish twist as he performed postmortems on bats, lizards, locusts, and newts. He dismembered them and interchanged their parts, using the results to create drawings of frightening creatures with which he could embellish shields and masks. Notably, he never abandoned this obsession with the hideous.

Though the nature of Leonardo's birth carried no terrible shame in his time, there were legal disabilities associated with his illegitimacy. Leonardo was prohibited from entering many professions, including that of a notary. Rights of inheritance were impaired, which later landed him in a legal battle with his legitimate half brothers over the estate of his uncle Francesco. In his early life, probably the greatest impact of Leonardo's illegitimate birth was in relation to his education. Most professions barred illegitimate sons, and so Leonardo was not given a

LEONARDO'S FATHER RECEIVES A SCARE

Vasari tells the story of a peasant employed by Leonardo's father, Ser Piero da Vinci, who came to Ser Piero with a buckler, or circular shield, he had made from a tree he had cut down. The peasant asked Ser Piero to take the buckler to Florence to be decorated by his talented son. Ser Piero, indebted to the man for his skillful assistance during hunting and fishing expeditions in the countryside, agreed to the request.

Leonardo decided to decorate the buckler with something that would produce "the same effect as once did the head of Medusa." He gathered a collection of grasshoppers, crickets, bats, serpents, and such, and using parts from each, put together a creature "most horrible and terrifying," represented on the buckler "belching forth venom from its open throat, fire from its eyes, and smoke from its nostrils."

When Leonardo's father came to pick up the buckler, Leonardo placed it opposite the entrance in such a way as to produce the most frightening effect on whomever entered the room. His father was startled and astounded, much to Leonardo's delight, and thought the buckler "nothing short of a miracle." He was so taken with it and its effect upon anyone seeing it for the first time that he returned a different buckler to the peasant and kept the one decorated by Leonardo, which he later sold for a hundred ducats—quite a considerable sum. Ser Piero's reaction was similar to that of today's moviegoers who are astounded by computer-enhanced special effects.

Opposite: Leonardo, ink drawing of a stream with a swan and duck, set against a backdrop of rugged cliffs, possibly the Balze Rocks in Valdarno, near Arezzo 1475-1480. Royal Collection, Windsor Castle, U.K.

humanistic, literary grammar school education in which he would have learned Latin, the primary intellectual language of his age. He took the shorter course of practical reading and writing in the vernacular and basic arithmetic that prepared him to be apprenticed in a trade. Almost nothing is known of the specifics of Leonardo's schooling—where he went to school or who were his teachers. Vasari touches on the subject of his education only briefly. He states that Leonardo was not necessarily the best student when he was learning his letters, because like the adult Leonardo he "set himself to learn many things, and then, after having begun them, abandoned them." We do not know Vasari's source for this information. He may be projecting his knowledge of the older Leonardo onto the child, for the behavior described was a defining characteristic of the adult. Vasari also tells us that Leonardo drove an arithmetic instructor to distraction with his pestering questions, that he loved music, learned to play the lyre, sang like an angel, and was always drawing and sculpting.

The most common track of schooling at the time for someone of Leonardo's prospects—on his way to apprenticeship in a trade—would have started with elementary school, where basic reading and writing in the vernacular were taught. At about age nine or ten a student moved on to a school known as an *abaco,* where the curriculum comprised the basic mathematics needed to carry on a business or trade. At age 12 or 13, a student was ready to become an apprentice in the shop or workplace of a master artisan.

In 1464, Ser Piero's wife, Albiera, died in childbirth, after 12 years of childless marriage. Within a year Piero remarried. Again he made an advantageous match in the world of the notaries, marrying Francesca Lanfredini, the daughter of Ser Giuliano Lanfredini, a powerful Florentine notary. Before long they took a house on the corner of the Piazza della Signoria, and around this same time in the mid-1460s, Leonardo's schooling ended, and he came to live with his father in Florence.

Above: Leonardo, "Bust of Condottiere in Profile," inspired by a lost Verrocchio relief of Darius, ca 1475-1480. Galleria degli Uffizi, Florence

Opposite: Giorgio Vasari, "Homage to Lorenzo 'the Magnificent,' " ca 1534. Sala di Lorenzo, Palazzo Vecchio, Florence

The Move to Florence

Leonardo's arrival in Florence coincided with a time of political upheaval and uncertainty. Cosimo de' Medici, the de facto ruler of Florence, who had done so much during his reign to enhance the cultural and artistic reputation and prominence of his city, died in 1464. He was succeeded by his son Piero, a weak and ineffective leader. Before long the city was divided into warring factions, the anti-Medici party headed by Luca Pitti, leader of another of the richest families in the city. In 1466 there was an attempted coup. Leaders of the insurgents were exiled and sought the help of outside powers, including Venice. Swords were drawn and shots were fired, but in the end Cosimo's son Lorenzo emerged from the fray and assumed power in 1469 at the age of 20. He proved to be a superb diplomat. Through his efforts a carefully calibrated balance of power preserved a tentative peace in Italy during most of the remainder of the century. Under this peace, trade

"Iron rusts from disuse;
stagnant water loses
its purity and in cold
weather becomes frozen;
even so does inaction sap
the vigor of the mind."
—LEONARDO

The Palazzo Vecchio (formerly named the
Palazzo della Signoria), with its prominent
tower, looms over the Piazza della Signoria.
Construction on the palazzo commenced in the
late 13th century, following the master plans of
Arnolfo di Cambio. At lower right, the Loggia
dei Lanzi features statuary executed by Benvenuto
Cellini and Giambologna.

ANDREA DEL VERROCCHIO

"Andrea del Verrocchio ... had a manner somewhat hard and crude, as one who acquired it rather by infinite study than by the facility of a natural gift."

—VASARI

Leonardo served his apprenticeship in the bottega, or workshop, of sculptor-goldsmith-painter Andrea del Verrocchio. Like his famous pupil, Verrocchio was born out of wedlock—as Andrea di Michele di Francesco de' Cioni; later, following a convention of the time, he adopted the name of his master, goldsmith Giuliano Verrocchio. It is also known that he trained alongside painter Filippo Lippi; that he frequently visited the workshop of Donatello; and that he may have even apprenticed with him, though this claim has been disputed. What is not disputed, however, is that when he ran his own studio in Florence, in addition to Leonardo, he trained a number of the finest young artists of Florence—Sandro Botticelli, Pietro Perugino, Lorenzo di Credi, and Domenico Ghirlandaio, who in turn would train Michelangelo. His most important patrons were the first family of Florence, the Medici.

Verrocchio's artistic legacy as a sculptor far surpasses his legacy as a painter. His masterpieces are a pair of monumental bronzes: "Christ and St. Thomas" and the equestrian statue of the condottiere Colleoni, which stands in Venice, where Verrocchio created it, after relocating his studio there in 1478. For the latter bronze, he designed and carved the clay model and created the mold required for the lost wax method—but died unexpectedly before the bronze could be poured. The casting was completed by the Venetian

Opposite: Verrocchio, "Tobias and the Angel," 1473. National Gallery, London

caster Alessandro Leopardi, who shamelessly inscribed his own name on the statue—a work which 19th-century Swiss historian Jacob Burkhardt described as "the finest equestrian statue ever made."

Smaller sculptures by Verrocchio include the regal bronze statue of "David" and the marble bust "Lady with a Bouquet of Flowers" (both in the Museo Nazionale del Bargello, Florence); and the busts of "Lorenzo the Magnificent" and "Guiliano de Medici" (both in the National Gallery of Art, Washington, D.C.). The subjects of Verrocchio's sculptures often display subtle smiles and carefully articulated hands, especially evident in "Lady with a Bouquet of Flowers." Verrocchio frequently impressed upon his apprentices the value of rendering expressiveness with the hands, and he urged them to build the body from the inside out. "Know the bones and muscles underlying the clothes," he told his pupils. These lessons resonated with Leonardo. This also had been an essential message in Leon Batista Alberti's treatise *On Painting*, which deeply influenced Leonardo, and probably Verrocchio before him.

In any modern discussion of Leonardo's early years, two works by Verrocchio rise to epic significance: "Tobias and the Angel," the depiction of a then popular apocryphal story

in which Tobias is asked by his father to collect a debt, and the archangel Raphael—believed by Florentines to be the guardian angel of travel—accompanies the boy; and "Baptism of Christ." According to curator David Brown of Washington's National Gallery of Art, Leonardo most likely painted the dog scampering underfoot in "Tobias and the Angel," as well as the angel in the lower left of the "Baptism of Christ."

Above: Andrea del Verrocchio, "Equestrian Statue of Condottiere Bartolomeo Colleoni," ca 1488. Venice

Below: Verrocchio. "Lady with a Bouquet of Flowers," 1475-1480. Museo Nazionale del Bargello, Florence

flourished and Italy's remarkable prosperity reached new levels. In Florence, more than in any other city—with the active encouragement of Lorenzo—that prosperity found an outlet in the arts, helping to push the already extraordinary output of her artists and architects to unprecedented heights.

It was in this confusing but exciting milieu that Leonardo began an apprenticeship intended to set him on his way to making a living. Vasari tells us that Ser Piero, having taken notice of the boy's constant pursuit of drawing and sculpting, took some of Leonardo's work to his friend Andrea del Verrocchio—a goldsmith, sculptor, painter, and all-around craftsman in the decorative arts—whose workshop on the Via Gibellina was just around the corner, and asked if the boy had any prospects. "Andrea was astonished to see the extraordinary beginnings of Leonardo," Vasari tells us, and he urged Ser Piero to have him apprenticed at once in the study of art. Leonardo was excited at the prospect. A deal was agreed to, and very soon thereafter, probably sometime in 1466, Leonardo became a student in the most important artistic bottega in Florence.

Verrocchio, his name translating as the "true eye," was the preeminent painter, sculptor, and goldsmith of his time, and he operated an unusually successful studio for apprentice artists. Here Leonardo met other young artists, among them Botticelli, Perugino (ca 1445-1523), and Lorenzo di Credi (ca 1459-1537). He learned the basic skills of mixing colors, casting bronzes, painting, and sculpting. And under the tutelage of Master Verrocchio he was introduced to perspective, chiaroscuro (use of light and shade for accentuating forms) and sfumato (a hazy, obscuring effect for creating a sense of distance)—artistic devices just being formulated to produce more realistic effects in painting. Leonardo not only used these devices but also had a major hand in their advancement.

Verrocchio's studio presented an ideal environment for a future engineer. Accordingly, when the studio received an order for a gilded copper sphere eight feet in diameter to crown the wondrous dome of the Cathedral of Santa Maria del Fiore, the 19-year-old Leonardo participated fully in the project. Verrocchio's team studied the pioneering technology that Brunelleschi had created for the original construction of the dome and succeeded in placing the ball gently and securely on top. And the young Leonardo, studying Brunelleschi's machinery, gained an abiding respect for innovation. Declaring mechanics "the paradise of all the sciences," he went on to design machines with parts that displayed intricate interlocking functions. Only in the late 20th century were some of these endeavors finally recognized as ruminations on robotics.

Verrocchio's frequent admonition to his apprentices to build the body from the inside out helped mold Leonardo's practice as a future scientist as well as an artist. But the master had no idea how far Leonardo would go in his pursuit of learning both human and animal anatomy. In two different periods of his life, Leonardo put everything aside to pursue the study of anatomy, even breaking the law to perform postmortems. In turn he cautioned his own followers, as well as future generations of scientists and artists, to learn from nature, not from each other. And to the scientists among them, he impressed the need to seek mathematical demonstration. In 1472 Leonardo, along with other members of Verrocchio's

workshop, including the master himself, joined the recently founded St. Luke's confraternity of painters in Florence. He could have struck out on his own at this time, but he preferred to remain in Verrocchio's workshop for another four years. From his earliest days as an apprentice, few works of art wholly by his own hand have survived. Although he collaborated with others on some paintings, the earliest extant work exclusively by Leonardo is thought to be an ink sketch of his beloved Arno Valley, dated August 5, 1473. It displays the impeccable perspective and the hazy air of receding distance—elements of the art and the science—that would distinguish all his works.

As an apprentice, Leonardo took on assignments executing the less significant portions of paintings awarded as commissions to Verrocchio. In his master's "Tobias and the Angel," Leonardo most likely painted the small dog, a white Bolognese terrier, as well as the curls in the hair of Tobias, and perhaps the fish in Tobias's hand. First pointed out by David Alan Brown, curator of Renaissance art at the National Gallery of Art, Washington, D.C., this may well be the very first example we have of Leonardo's brushwork. The magically diaphanous dog is bouncing alongside the legs of the main characters, who are themselves rather stiff and motionless. Both the dog's wavy hair and the left-handed brushstrokes in the locks above the ear of Tobias are signature Leonardo.

The First Portrait

Between 1473 and 1475, Leonardo undertook the assignment of painting the portrait of Ginevra de' Benci, daughter of the Florentine banker Amerigo de' Benci. The circumstances of how he received the commission are not known, but a reasonable guess would be that Verrocchio, already exceedingly famous and very much in demand, simply did not have the time to do the portrait himself. Moreover, since Verrocchio preferred to do sculpture, a craft in which he especially excelled, he probably passed the commission on to his promising young apprentice.

Above: Leonardo, "Study of Hands," possibly for the "Ginevra," 1474. Metalpoint on paper. Royal Collection, Windsor Castle, U.K.

Opposite: Leonardo, "Ginevra de' Benci," ca 1474. National Gallery of Art, Washington, D.C.

The portrait of Ginevra appears on a nearly square wooden panel that Leonardo painted on both sides. On the front of the panel is the portrait of the girl standing before a juniper bush (*ginepro* in Italian, a play on the subject's name) with the Tuscan countryside rolling out into the distance. Ginevra's face is pale, reflecting the fact that noble ladies did not spend much time outdoors and did not acquire a tan. Her hairline has been plucked, accentuating the smooth dome of the forehead, a sign of wisdom; and in the distant landscape one can make out a pair of church spires, an emblem of piety. Leonardo has imbued his 16-year-old subject's face with a touching sadness; she will soon marry (or has recently married) a widower twice her age, the prominent magistrate Luigi Nicollini.

Painting the reverse side of the wooden panel had a salutary effect, since it has kept the panel perfectly flat for over 500 years. Warping is caused by the cells on one side of a wooden panel absorbing more moisture than cells on the other. This panel was hermetically sealed; however, even that cannot stop discoloration of the painting's protective varnish, which yellows over the years. In 1992 the painting underwent thorough cleaning and restoration, and what emerged was even more beautiful than expected. Viewing the painting through this filter of yellow varnish converted the blues to green and reds to orange. Removing the yellowed varnish from the portrait restored Leonardo's original blue hues in the sky. It also revealed the rich tones Leonardo originally used in painting the girl's aristocratic complexion, much paler and more porcelain-like than before the restoration.

The "Ginevra de' Benci," which hangs in Washington's National Gallery of Art, is Leonardo's only painting exhibited outside Europe. Leonardo definitively painted two other secular portraits of women during his lifetime: The "Lady with an Ermine," produced about 15 years later, in Milan, and the "Mona Lisa," another 15 years later still, after his return to Florence. A third one, "La Belle Ferronière," painted circa 1495, is attributed to him, but its provenance remains in question.

HIGH-TECH INVESTIGATION OF "GINEVRA"

"Leonardo's 'Ginevra' is emblematic of this transition toward a more naturalistic and expressive rendering of the sitter as an individual."

—DAVID ALAN BROWN

LEONARDO DA VINCI, ORIGINS OF GENIUS, 1998

Modern restoration and conservation of valuable paintings frequently call for the use of technology that Leonardo himself would have found fascinating. In his investigation in the 1980s and '90s of the "Ginevra de' Benci," conservator David Bull of the National Gallery of Art, Washington, D.C., relied mainly on the techniques of x-radiography, infrared reflectography, and stereoscopic microscopy. With x-radiography, x-rays are used to probe deep into a painting in order to ascertain the beginnings and the structure of the work. Leonardo painted the portrait of Ginevra on one side of a poplar board, and an altogether different painting on the reverse side. The framed double-sided work, mounted on a vertical plinth in a prime location in the gallery, allows viewing of both sides. With x-radiography applied to the "Ginevra," the images on the two sides of the panel are compressed, making both sides visible simultaneously. This technique also reveals overpainting and any previous images.

On the reverse side of the "Ginevra," underlying the Benci family motto "Beauty adorns Virtue" is another, "Virtue and Honor" (the motif of Bernardo Bembo, who was the Venetian Ambassador to Florence, became Ginevra's platonic lover, and is perhaps the person who commissioned Leonardo to do the painting). Thus Ginevra's own motto was

Opposite: "Ginevra de' Benci Restored," executed by the staff of the National Gallery of Art, Washington, D.C., using the Windsor "Study of Hands"

Above: The reverse side of the portrait, as restored by the staff of the National Gallery of Art

overpainted on Bembo's. This gives rise to controversy about the date of this painting. Was it commissioned in 1473-74 by her husband as a wedding present, as had generally been believed, or, as others now suggest, was it commissioned by Bembo after his arrival in Florence as ambassador in January 1475? These details matter only in placing Leonardo's paintings in the correct chronological order.

GINEVRA'S MISSING HANDS

The portrait of Ginevra is on a poplar panel with an almost square shape, but was it always a square? A square format was rare for paintings in the Renaissance. Is there a missing part of the painting? In the case of the Ginevra, two of its edges, the right and the bottom on the obverse side, display evidence of damage and of having been sawed.

Accepting the fact that Leonardo considered the hands as important as the face for conveying emotion, Dr. Brown and members of the technical staff of the gallery digitally conjoined the portrait of Ginevra and a study of hands drawn by Leonardo that now resides in the Queen of England's collection. The effect was nothing short of extraordinary.

The clue to how the hands and torso should be aligned was found on the reverse side of the portrait, where the vertically configured juniper twig is offset by about half an inch in the direction of the damaged edge. Operating under the hypothesis that the juniper twig had been aligned at the center of the original panel, the first step was to add half an inch to the damaged edge and then fill in the space digitally. The branches embracing the juniper were extended downward to converge near the bottom of the panel. Then on the obverse side, the hands were digitally overlaid at the bottom of Ginevra's image, and manipulated—translated, rotated, and re-scaled—until the artist's style and intent were achieved. The digital colorization process leaves the colors of the hands and dress somewhat flat. Moreover, distinctly missing, especially in the hands, are the softness and grace that Leonardo would normally have captured in an oil portrait. We do, nonetheless, have an opportunity to see Ginevra's position as it might have appeared in the original portrait.

LEONARDO LEAVES HIS FINGERPRINTS

In 1992 the National Gallery of Art gave its single most important holding, Leonardo's "Ginevra de' Benci," a thorough cleaning. In the process, the conservator, while examining the painting with a stereoscopic microscope, discovered Leonardo's fingerprints. They are located in the juniper bushes to the right of Ginevra's neck. When the "Portrait of Cecilia Gallerani" visited the United States that same year, the conservation department of the National Gallery of Art found Leonardo's fingerprints on that portrait as well. They are on the neck of the subject. This is entirely compatible with the most recent confirmation from ion beam studies that Leonardo was using the technique of velature: He had abandoned the use of a palette to mix his colors, instead applying unadulterated colors with his fingers in thin layers one on top of the other directly on the panel. The National Gallery laboratory, although equipped with the latest high-tech equipment for probing paintings, lacked the specialized camera to photograph the fingerprints. Accordingly, the staff contacted the FBI laboratories to photograph Leonardo's fingerprints. They are now on file with the FBI.

Competing with the "Ginevra" for the distinction of being the earliest painting exclusively by Leonardo is the "Annunciation." This work exists in two versions—one in the Uffizi in Florence, the other in the Louvre in Paris—and controversy has surrounded their attribution. The early works of Leonardo have often been mistakenly attributed to Lorenzo di Credi and vice versa, and sometimes the attributions have been reversed after careful scrutiny. The source of the confusion lies in the common training the two artists received as fellow apprentices in Verrocchio's workshop. The general consensus among art scholars now is that the Uffizi version of the "Annunciation," executed around 1472-75, is more likely by Leonardo. The angel's sleeves

are a virtual facsimile of the sleeves in a known Leonardo drawing, both almost diaphanous in their appearance—exhibiting a Leonardesque quality. Art scholars are able to date the painting through the design on the sarcophagus positioned between the angel and the Virgin. (Verrocchio had carved the sarcophagus for the recently deceased Piero de' Medici, who died in 1469.) The Louvre "Annunciation," painted around 1478-1485 on a commission originally awarded to Verrocchio by the church authorities of the Cathedral of Pistoia, displays in the draperies characteristic touches of Lorenzo di Credi.

The paintings depict the moment when the archangel Gabriel appears before the Virgin Mary to announce the imminent arrival of Christ. In their compositions the two paintings resemble each other. The setting, an enclosed garden, and the Madonna lily held by the messenger both symbolize virginity. In the Louvre version, the building to the right displays a simple flat façade; in the larger and more powerful Uffizi version, the building has prominent white quoins, deeply beveled stones along the corners of the building, and a pair of walls forming an inside corner (very difficult to see in most reproductions). The painting contains minor mistakes in perspective. The three cypress trees in the background on the left should show diminishing size in receding to the left. Also the lectern on the table appears to be closer than the Virgin, causing Leonardo to elongate the Virgin's arm in order to reach the book on the lectern. Nonetheless, the "Annunciation" is a masterpiece, and, in the words of David Alan Brown, "the work of an immensely gifted artist who was still immature."

Even after displaying extraordinary skills in painting the "Ginevra" and the "Annunciation," Leonardo returned to painting minor parts of Verrocchio's own commissions. In one of the master's more important works, the "Baptism of Christ" (ca 1475), Leonardo was assigned the task of painting one of two angels in a lower

corner. This proved a strategic mistake for Verrocchio, who upon glimpsing the young apprentice's work, put down the brush and vowed never to paint again. Leonardo's angel, kneeling in the lower left corner, becomes the focal point of the entire painting, detracting from the much larger figures of Christ and John the Baptist painted by Verrocchio himself. This story becomes, in Vasari's *Lives*, a parable of emerging genius—an example of the pupil transcending the master. Even Leonardo himself would write: "It would be a poor student who would not surpass his master."

One cannot miss the slight helical twist in the pose of Leonardo's angel or the spirals in the curls of the hair, the very same spirals that Leonardo had employed in the "Ginevra," and that he later used in his drawings of the vortices in turbulent water; the blood rushing out through a heart valve, causing it to close; and wind shear during a violent thunderstorm. The conjoining of art, science, and mathematics—and an uncanny ability to observe, record, and make connections—were already on display in such an early work.

Above: Lorenzo di Credi, "Annunciation," ca 1478-1485. Oil on wood. Musée du Louvre, Paris

Opposite: Leonardo, "Annunciation," ca 1472-75. Galleria degli Uffizi, Florence

Leonardo's Bottega

In 1476 Leonardo finally opened his own small workshop, and in the next four to five years worked on a number of private commissions, all religious in nature— "Madonna with a Flower," known as the "Benois Madonna" (Hermitage, St. Petersburg); "Madonna with the Carnation" (Alte Pinakothek, Munich); "Madonna of the Yarnwinder" (one version of which was recently stolen from a private collection in Scotland); and the "Madonna Litta" (Hermitage). Another painting whose provenance was in doubt, the "Madonna of the Pomegranate," was recently attributed to a collaborative effort of Verrocchio and di Credi.

In this period, Leonardo began work on a large-scale "Adoration of the Magi," an altarpiece for the friars of the Augustinian monastery San Donato a Scopeto. Leonardo, with his personal charm, was a good diplomat, but not a very astute businessman. He always got his foot in the door, but did not always secure the best possible deal for himself—and this project turned out to be troubled from beginning to end.

The contract for the March 1481 commission was extraordinarily complex. Leonardo was to provide all the paints and the gold that would be used in the work; a tight time limit of 30 months was set for the completion of the painting. There would be no payment in advance; and after the painting was turned over to

Above: Leonardo, "Perspectival Study for the 'Adoration of the Magi,'" ca 1480. Galleria degli Uffizi, Florence

Opposite: Leonardo, "Adoration of the Magi," unfinished, ca 1481. Tempera mixed with oil on wood. Galleria degli Uffizi, Florence. The shepherd boy (at far right) is believed to be a self-portrait of the young Leonardo.

the monks, the artist would be paid, not in cash, but in land. He would receive one-third of a piece of property that the church had received as a bequest from a wealthy patron. Once the work was completed and he received title to the land, he was not to sell or otherwise encumber it in any way for three full years. And indeed, if he desired to sell it, he could only sell it back to the church, and the price of this sale was set at 300 florins. Finally, a penalty clause in the contract stipulated that if he did not finish the work on time, he would relinquish whatever of the work he had done so far and forfeit any claim on any payment owed. The conditions clearly favored the friars, but Leonardo needed the commission. Ironically, it was likely that his father, Ser Piero, had participated in negotiating the contract, since he had been the business agent for the monastery since 1479.

Leonardo, in the most famous of his studies (above) leading up to the work itself, conceived it as a large rectangular panel—effectively a stage with a palace in ruins as the backdrop. A set of stairs rises diagonally to an upper floor of the structure. The stage is teeming with life—people of all ages, animals, men on rearing horses, a recumbent camel (unusual even for Leonardo). In this and other early studies for the "Adoration," the figures, even those on horseback, are all naked. His notion of building the body from the inside out—from skeleton and organs to flesh—even extended to the next layer, the clothing of the bodies. Dressed, they had to move as naturally as they did in the nude. Carefully composed lines of perspective all converge at a distant vanishing point on the horizon line.

By the time he undertook the actual application of paint to panel, Leonardo had settled on a nearly square work. Some of the animals he had conceived of in the early drawing are still on view in the painting. The Virgin and the infant Christ are center stage with the attendant sup-plicants, mostly in prone positions as they approach the divine mother and son. The mélange of figures is virtually a puzzle of hidden images. A total of 66 individuals have been counted. The stairs to the upper ramparts are as in the drawing, although the palace in ruins is in a more dilapidated state. Symbolism is rife. The tree of life behind the Virgin and the infant Christ, symbolic of emerging Christianity, stands in stark contrast to the ruins, symbolic of declining Greek and Roman paganism. But at a different level, the roots of the tree of life rising from the infant Christ are reflections of the Prophet Isaiah's pronouncement: "There shall come forth a rod of the stem of Jesse, and a branch shall grow out of his roots." The supplicants appear ghoulish, fearful, and threatening at the same time. Biographer Charles Nicholl asks the question, "Where is Joseph?" and sees a recurring motif in this and future Leonardo depictions of the holy family:

Leonardo always excises Joseph from the holy family. He is missing from the 'Virgin of the Rocks' (which narratively takes place during the flight from Egypt, and so ought to include him), and he is missing from the various versions of the 'Virgin and Child with St. Anne,' where the third member of the family is not the child's father but his grandmother. One does not have to be a Freudian to feel that there are deep psychological currents here.

In the lower right-hand corner stands a lone shepherd, staff in hand, distracted, gazing away furtively. Art historians have long speculated that this figure might just be

"Alas this man will never do anything, because he is already thinking of the end before he has even begun the work."

—POPE LEO X

a self-portrait of the young artist himself, and there is indeed a compelling likeness to the "David" of Verrocchio, widely considered to be modeled on Leonardo. Moreover, an artist inserting himself into his own paintings was common in the Renaissance.

Along with the studies leading up to the work, this painting holds immense significance as a window into Leonardo's manner of working. Leonardo's plan for the painting, his experimentation with compositional elements and media, and his mathematical musings—in this instance, the exquisite one-point perspective—are all apparent, along with the psychological elements in the characters, including the beasts. The animals appear to convey human emotions and reactions. But as is so often the case with Leonardo, the conception was more important than the execution. He tired of the "Adoration," much to the chagrin of his patron friars, and moved on to the next project. Perhaps in drawing up the original contract, the friars had been aware of the artist's growing predilection for delay and procrastination.

When it became clear to the friars of San Donato a Scopeto that Leonardo had completely lost interest in finishing the work, another famous artist, Fra Filippo Lippi, was invited from Milan to paint an alternative altarpiece that would be completed six years later.

Leonardo's unfinished work, one of the enduring gems of the Uffizi, was recently shown to bear the crude brushwork of a painter distinctly less talented, who worked on the painting perhaps a century after Leonardo had abandoned it. The shape of the Virgin's feet is among the elements of the overpainting that reveal the contamination of the lesser talent. The underpainting, however, is Leonardo—but early Leonardo—just beginning to slough off the Verrocchio style on the path to developing his own. What Leonardo presents in the "Adoration"—"a certain dreamlike quality with an edge of foreboding," according to critic Maxine Anabell, is a preview of the great works to come.

The "Adoration of the Magi" was followed by another unfinished painting, "St. Jerome." Leonardo in this period of desperate frustration worked on the painting of St. Jerome sometime between 1480 and 1482. He would certainly have felt a kinship with St. Jerome, the patron saint of nature and the wilderness. The painting became a vehicle for him partly to commiserate with and partly to commune with the stoic holy man. Jerome has been described as "haggard from fasting and penitence; at the same time his eyes display determination and will-power." He is seen with a rock in hand, getting ready to beat on his own breast. For Leonardo, St. Jerome's imminent act of penance perfectly mirrors his own mood of melancholy at a time of mounting personal crises. Annotations in the margins of his notebooks include poignant entries, "Why do you suffer so? The greater one is, the greater grows one's capacity for suffering. I thought I was learning to live; I was only learning to die."

In the painting of St. Jerome, Leonardo shows his faultless knowledge of human anatomy. The muscles of the sunken face, the muscles of the arms and shoulders, the flesh and skin masterfully conjoined to the underlying skeleton are all stretched to reveal the tendons. This is anatomical drawing at its very best, far surpassing Master Verrocchio's depiction of the gaunt figures in the "Baptism of Christ." Moreover, in contrast to the plastic visages portrayed by Verrocchio, "St. Jerome" resonates with psychological overtones, prefiguring the tension in the "Last Supper."

Troubling Times

Thinking through a problem, planning a composition, or concocting a new and better paint or varnish were exercises far more exciting to Leonardo than laboring to finish a project. Art historians have often suggested that the reason Leonardo produced so few works and left so many unfinished was that painting was just too easy for him. But no other artist appears to have agonized more than Leonardo in conceiving and planning a work of art. It was only the execution phase that failed to engage his lasting attention, or may have even bored him. Patience and perseverance were virtues Leonardo rarely displayed. As a man of almost 30, he was revealing the trait of abandoning one interest for another, a trait that Vasari would attribute to him as a young schoolboy. Leonardo's failure to deliver these two commissions on time reinforced a troubling pattern in his work routine—flitting from one unfinished project to another.

Opposite: Leonardo, "St. Jerome," ca 1481. Vatican Museums

Following pages: One of several bridges spanning the Arno River, the Ponte Vecchio— "old bridge"—was already more than a century old when Leonardo was born.

GOLDEN RECTANGLE

Leonardo was obsessed with mathematics and its application to fields seemingly far removed from the quantitative—including art. His pronouncement, "Let no man who is not a mathematician read the elements of my work," reveals his conviction that math underlies all fields worthy of his attention.

The golden rectangle, defined by a length-to-width ratio of 1:1.618... (a number designated by the Greek letter *phi, ϕ*), is associated with the "divine proportion," known since the Pythagoreans of the sixth century B.C. In antiquity the shape appears in several defining works of architecture, including the façade of the Parthenon. The letter ϕ honors the surpassing Greek sculptor-architect Phidias, who had collaborated in the design of the great edifice and carved its statuary.

Leonardo, who would later in his life collaborate with Luca Pacioli on a mathematical treatise, *De Divina Proportione*, was familiar with the golden rectangle, as well as with regular and semiregular polyhedral figures that he depicted in rough sketches in his notebooks and presented formally in the treatise. In his unfinished "St. Jerome," the figure of the ascetic is framed exactly by the superimposed golden rectangle. In light of Leonardo's obsession with numbers and patterns, it is most likely not a coincidence for him to have inscribed the figure of St. Jerome within a golden rectangle. He also imbued the portraits of young women, including the "Mona Lisa," with elements of the mathematics of aesthetics.

THE THREE DAVIDS

The Book of Samuel in the Old Testament records a battle between the Israelites and the Philistines at Socoh in Judea. Before the warriors have actually engaged in combat, however, a colossal Philistine named Goliath emerges from the camp and taunts the Israelites, challenging them to send out a warrior to engage him in single combat. To the startled surprise of the Israelites and their king, Saul, a slight youth, David, volunteers to take on Goliath. Saul offers David his own sword and armor, but is alarmed to see the frail young man refuse the offer. David steps up, armed only with his sling and five rocks that he has selected from a nearby riverbed. Calmly facing the heavily armored Goliath, David places one of the rocks in his sling, whirls it around, and launches it at his foe. The rock strikes the Philistine between the eyes. Stunned, he topples over face down. David approaches the prone giant, picks up his sword, and with one bold swing, cuts off his head. The shocked Philistine army retreats in panic; the Israelites rejoice.

Florence, in a continuous state of conflict with one powerful neighbor or another, saw itself as the "David of northern Italy." Accordingly, whenever a skirmish concluded with their city's survival, Florentines celebrated by commissioning a statue of their idol. Many representations of David were created. One of the three most important is a bronze by Donatello, which melds elements of the classical ideal and the emerging Italian reality. This David is a frail adolescent with underdeveloped musculature. He has won the battle and is standing over the giant, holding the Philistine's oversize sword, the point resting on the ground. His face expresses both joy and incredulity at his accomplishment. A revolutionary statue, it is a free-standing male nude, the first since antiquity.

The second memorable "David" is by Verrocchio, commissioned by the Medici. This David—a slightly older youth than Donatello's—is clothed and also stands victorious after combat. He holds a small sword, resembling a large dagger. The expression on Verrocchio's "David" exudes pride and confidence, and it bears that slight smile, a trademark of Verrocchio's sculpture. Cast in bronze, it measures 50 inches in height and is a gem of the Renaissance.

The third "David" is by Michelangelo. It is a colossal marble carving 13.5 feet tall, three times the height of Verrocchio's "David," portraying an older David standing contrapposto (with the plane of the chest at a different angle than that of the hips). Unlike the other two, this David is a young muscled Adonis, bristling with confidence. Against this David, Goliath has no chance. The leather strap of his sling rests on his shoulder; his massive right hand hangs freely at his side. The tension of the moment is conveyed by the intensity of his gaze and his tightened muscles. This is the instant when he has decided to act, to launch the rock. Michelangelo's "David" is the defining statue in Florence's David series—and laid to rest any future competition.

Above left: Donatello, "David," ca 1444. Muzeo Nazionale del Bargello, Florence

Above right: Andrea del Verrocchio, "David," 1473. Muzeo Nazionale del Bargello, Florence

Opposite: Michelangelo, "David," 1504. Galleria dell'Accademia, Florence

The "Annunciation" and "St. Jerome" are today, even in their unfinished condition, regarded as towering works of art, central to Leonardo's reputation. But in his own time their incomplete state detracted from that reputation. Patrons were paying for completed works, and Leonardo was not supplying them. In 1481 Lorenzo de' Medici was asked by the papacy to choose the best Florentine artists to help complete the decoration of the walls of the Sistine Chapel in Rome. This commission was diplomatically important as well as artistically prestigious. Lorenzo was seeking more amicable relations with the Vatican and the Papal States, and so he sent Botticelli, Ghirlandaio, Perugino, and Cosimo Rosselli—the best of Florence—to paint ten

Botticelli, "Seventh Circle of Hell," an illustration for Dante's Divine Comedy, *showing the torments of the homosexuals*

scenes from the Bible on the walls of the famous chapel. Leonardo's name was not on the list. He may have been excluded simply because he could no longer be relied upon to finish anything. It may also have been that he had fallen out of favor with the Medici. Whatever the reason for Lorenzo's failure to endorse him for the mission to Rome, the humiliation that Leonardo surely felt was only the latest in a series of troubles that had plagued him from about the time he had left Verrocchio's workshop to start his own. And with commissions drying up and the most significant names in the business heading to Rome, Leonardo was beginning to feel isolated and must have sensed that the promise had gone out of his life in Florence.

His troubles had begun four years earlier. On April 8, 1476, an incident took place that must have devastated him personally. An anonymous denunciation, or *denuncia,* had been dropped into one of the *tamburi,* wooden boxes placed around the city to receive just such anonymous comments and denunciations from the public. The denuncia accused Leonardo and four other young men of participating in immoral activities and sinful acts, including sodomy, with a 17-year-old named Jacopo Salterelli. Who this Jacopo Salterelli was remains obscure. From the denuncia we know that he had a brother named Giovanni and they lived together in a goldsmith's shop, where they may both have been apprentices. What the motives might have been behind the anonymous denunciation are even more obscure—personal jealousies? a nosy, meddling neighbor? the commercial rivalry of a competing goldsmith? The trail ends almost before it begins. Two months later the case was dismissed, probably for political reasons. It seems one of the accused, Leonardo Tournabuoni, was a relative of Lorenzo de' Medici's mother. It would not do to have the leading family of the city dragged into an ongoing sex scandal. Discreet signals were no doubt given, and the legal case quickly collapsed. But the damage to the public character of the accused was certainly longer lasting.

In Renaissance Florence homosexuality was at least as common as anywhere else, and some prominent names among Florentines of the time were certainly gay. The sculptor Donatello was gay and Botticelli was generally believed to be, and had himself been the victim of a denuncia. Though homosexuality was openly tolerated in elite intellectual and artistic circles, flagrant public behavior or indifference to public opinion was not advisable. The church, of course, stridently disapproved. In Dante's *Inferno* homosexuals were consigned to the seventh circle of hell to wander forever in a burning desert. Preachers denounced the practice regularly. Florence was only a few years away from the social upheaval that would come in the wake of the sermonizing of the savage, puritanical moral reformer Savonarola. Sodomy was a capital crime—though the death penalty in such cases was rare—and accusations were to be taken seriously. A guilty verdict could result in the accused being heavily fined or exiled from the city. About two years after the affair of Jacopo Salterelli another such case arose, again involving Leonardo, and this time resulting in just such an exile, though not of Leonardo.

In February 1479, Giovanni Bentivoglio, paramount leader of the city of Bologna, wrote a letter to Lorenzo de' Medici of Florence on behalf of a young man named "Paulo de Leonardo de Vinci da Firenze." Paulo apparently had been exiled to Bologna from Florence, probably a year or so before, because of the dissolute, wicked, and sinful life he had led there, the innuendo of the letter referring certainly to homosexuality. Lorenzo had requested that he be put in jail. Bentivoglio had complied with that wish. But now Paulo had learned a new trade, reformed his life, was living in a respectable manner, and would like to come back to Florence. The letter is a plea for Lorenzo's pardon and permission for the boy to return. The "de Leonardo de Vinci" indicates that he had been an apprentice in Leonardo's new bottega in Florence. As Verrocchio, Leonardo's master, had done, Paulo followed the convention of an apprentice taking the name of his master. The letter refers also to the "bad company" that Paulo had kept before his removal to Bologna. Even more so than in the earlier Salterelli accusation, specifics are vague and innuendo is the preferred mode of discussion. But something troubling to the authorities was definitely going on in Leonardo's workshop. And it had to have been going on, even if not with Leonardo's participation, at least with his

"There is no record of any woman in his life—not even a female friendship. On the other hand, he was soon surrounding himself with a constantly renewed court of remarkably beautiful young men."

—SERGE BRAMLY
LEONARDO : THE ARTIST AND THE MAN, 1995

knowledge. Did Lorenzo send the boy away to Bologna and jail under a cloud of blame to protect the rising reputation of Leonardo and to foster the appearance that he was keeping Florence and her artists free from the contamination of vice? Florence was gaining a reputation elsewhere as a haven for homosexuality. The Germans were beginning to use the word *Florenzer* as a synonym for sodomite. Or did he act decisively in order to prevent a gossipy reemergence of the Salterelli scandal? Whatever Lorenzo's motives may have been in choosing to resolve the issue as he did and thus save Leonardo's skin for a second time, his patience with Leonardo had to be wearing thin, and Leonardo's public persona must have been badly damaged.

Most scholars acknowledge that there seems little doubt that Leonardo was homosexual. Though he escaped any legal consequences stemming from the two episodes in Florence, their implications cannot be dismissed. When later facts and impressions are added to these early incidents the preponderance of the evidence becomes more convincing. Other young men will enter his life, most notably Giacomo Caprotti, to whom he gives the name "Salai," and to whom, we know from his own commentary, he becomes very attached. Some of his drawings are frankly homoerotic. And there are no known liaisons with or passionate attachments to women. Nineteenth-century art historians, particularly British commentators such as Walter Pater, immersed the question of Leonardo's sexuality in a shimmering haze of Platonic generalities, as they did most questions on the subject of sex. But beginning with an important essay by Sigmund Freud, modern students of Leonardo have come to accept and interpret Leonardo as a homosexual.

A Relentlessly Curious Mind

There is no evidence to lead us to believe that Leonardo was particularly comfortable with his sexuality. Sigmund Freud's essay of 1910, "Leonardo da Vinci: A Study in Psychosexuality," makes the persuasive argument that Leonardo was not at all comfortable with his sexuality or with sex in general, and that this helps explain much about the preoccupations, interests, and behavior of the man. There is no evidence for most of what Freud assumes, and anyway it is not the origins of Leonardo's condition that interest us but the consequences—and Freud argues that they were profound.

After the episodes in Florence, Leonardo stays clear of scandal. It is not likely that he had many, if any, regular sexual partners after he left town. Even the later relationship with Salai cannot be shown to have been a physical one. Freud makes the case that Leonardo's sexual energies were "sublimated … into curiosity, and … the powerful investigation impulse." The arc of Leonardo's career following his first years in Florence suggests this might be true. It is after he leaves Florence and his troubles behind to start a new life in Milan that Leonardo becomes most fully alive as the man whom we associate with art historian Kenneth Clark's famous remark about Leonardo being "the most relentlessly curious man in history."

In early 1481, that Leonardo was still in the future. By this time, the reality of life in Florence must have been oppressive for Leonardo. All his demons were in hot pursuit. His career was on hold. There were no commissions. Projects lay unfinished. A cloud hung over his lifestyle. And he seemed to have lost the favor of Lorenzo de' Medici. The Florentine sojourn was at a dead end.

Various possibilities have been proposed as the ostensible reason for his departure for Milan, probably in late 1481, but the dispirited and always restless Leonardo was surely ready for a change of venue. Whether he was sent by Lorenzo as a sort of cultural ambassador as some have suggested, maybe to get Leonardo finally out his hair, or came at the invitation of Ludovico Sforza, strongman of the city, or was responding to the possibility of winning the commission to create an equestrian statue of Ludovico's father, Francesco Sforza, or some combination of these or other reasons, we do not know. We do know that he left on the nearly 200-mile journey to Milan filled with ambitions, a sense of promise, and hopes for a fresh start. Eighteen years would pass before he again laid eyes on Florence.

Leonardo, "Madonna Litta," 1489-1491. Hermitage, St. Petersburg

"Although nature commences with reason and ends in experience it is necessary for us to do the opposite, that is to commence with experience and from this proceed to investigate the reason."

—LEONARDO

Leonardo's Sketchbook

Whether doing art or science, Leonardo was always seeking connections, having come to the conclusion early on that the most fertile ideas came from the synergy involved in exploring different intellectual worlds. He wrote: "Principles for the Development of a Complete Mind: Study the science of art. Study the art of science. Develop your senses—especially learn how to see. Realize that everything connects to everything else."

"Star of Bethlehem, Wood Anemone and Sun Spurge," 1505-1510. Royal Collection, Windsor Castle, U.K.

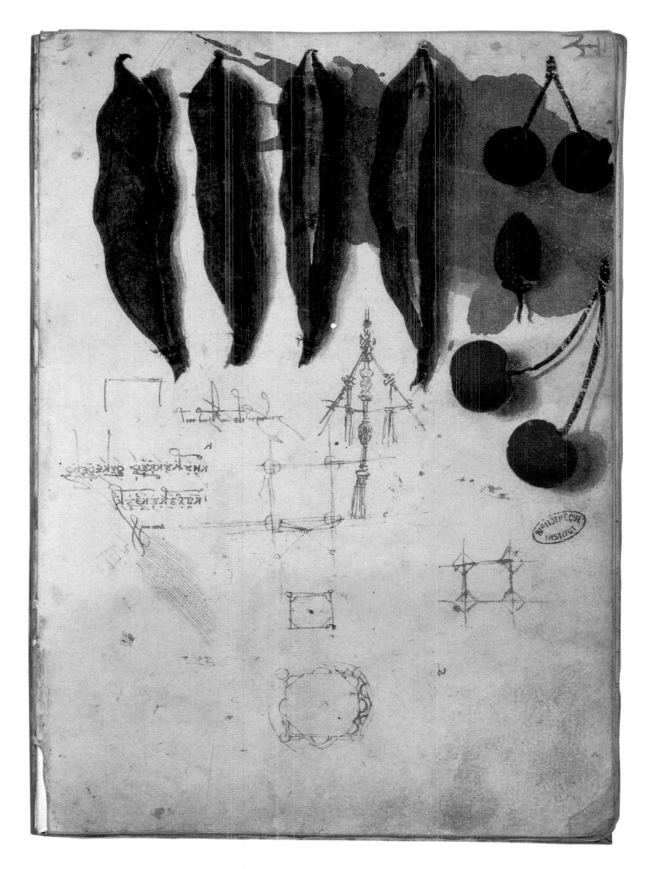

"Fruits and Vegetables," ca. 1487-1490. Paris Ms B, Institut de France, Paris

"A Study of the Fall of Light on a Face," ca 1499. Royal Collection, Windsor Castle, U.K.

"A Study of a Rearing Horse," 150⁴. Royal Collection, Windsor Castle, U.K.

"A Storm Over a Town," 1517-18. Royal Collection, Windsor Castle, U.K.

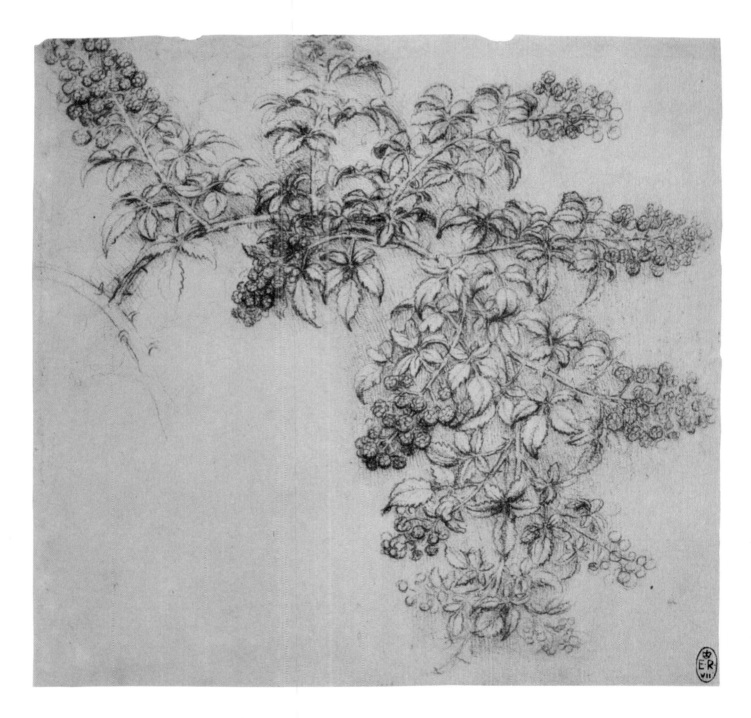

"Blackberries," 1505-1510. Royal Collection, Windsor Castle, U.K.

Chapter Three

> *"He who does not know the supreme certainty of mathematics is wallowing in confusion."*
>
> —LEONARDO

1481-1490
THE MILAN YEARS

1481-82 Leonardo moves to Milan.

1483 Leonardo begins first version of "Virgin of the Rocks."

1484-85 Bubonic plague ravages Milan.

1485 Leonardo begins studies on manned flight and on anatomy.

ca 1487 Leonardo enters architecture competition to build a dome on Milan's Duomo. He proposes plan for ideal double-tiered city.

1488 Verrocchio dies in Venice.

ca 1488-1490 Leonardo paints "Portrait of a Musician."

late 1480s Leonardo undertakes "Portrait of Cecilia Gallerani," aka "Lady with an Ermine."

1489 Leonardo begins work on the Sforza horse, a statue that will occupy him for six years.

Like so much else in his life, the date of Leonardo's departure from Florence is not known. It must have been after September 1481, the last record of his residence in Florence, and before April 1483, the first record that places him in Milan, where he signed a contract for painting the "Virgin of the Rocks." He most likely arrived sometime in early 1482; that date would certainly have given him time to settle in to his new surroundings and secure a commission as important as the "Virgin of the Rocks."

The Sforza Court

Milan in 1482 was an up-and-coming city, vying to keep up with Florence, Genoa, and Venice in the world of Italian city-state politics, and Ludovico Sforza was its vain, ambitious new leader. He had great plans for enhancing not only Milan's political power but also its cultural profile. Sforza, nicknamed "Il Moro" (the Moor)—because of his swarthy skin and possibly as a play on "Mauro," one of his given names—was a warlord; his family had gained its position in Milan by military means, and he certainly required military vigilance to maintain it. Ludovico was the fourth son of Francesco Sforza, the strongman who had established the rule of the Sforza in Milan in 1450 by displacing the Visconti, the previous ruling family, and declaring himself Duke. Francesco died in 1466 and was succeeded as duke by his son Galeazzo Maria, Ludovico's older brother, who had an execrable reputation for ruthlessness and cruelty; he was guilty of crimes, according to one Milanese historian, too shameful to even record. Galeazzo had enlarged, fortified, and improved Castello San Giovio, on the northern periphery of the city, making it the Castello Sforzesco, center of power for the Sforza. In the late 1480s, Corte Vecchia (once the Castello Visconti, home of Milan's previous ruling family) became Leonardo's home, studio, library, and study—a place where he would do some of his most dazzling artistic and intellectual work.

The hated Galeazzo was assassinated in 1476—and succeeded as duke by his ten-year-old son, Gian Galeazzo. Ludovico, however, became the effective

Opposite: Leonardo, "Virgin of the Rocks," second version 1495-99. National Gallery, London
Previous pages: Milan's Piazza del Duomo remains a center of activity, just as in Leonardo's time.

A map of Milan by Antonio Lafreri, dated 1573, shows the walls ringing the city, with the Castello Sforzesco at top center.

ruler, quickly securing the position of regent after his brother's demise. He hoped to redeem the reputation his brother had bequeathed to Milan and the Sforza dynasty and help Milan join Venice and Florence at the forefront of Italian culture. Two important projects had already jelled in his imagination—a new dome for the cathedral, which might be a respectable rival to Brunelleschi's achievement in Florence, and an equestrian statue of his father, Francesco, to be the largest such monument ever created. Leonardo had come to the right place at the right time.

According to biographer Giorgio Vasari, the powerful Sforza had actually invited Leonardo to Milan. Sforza loved *lira* (lyre) music and the talented

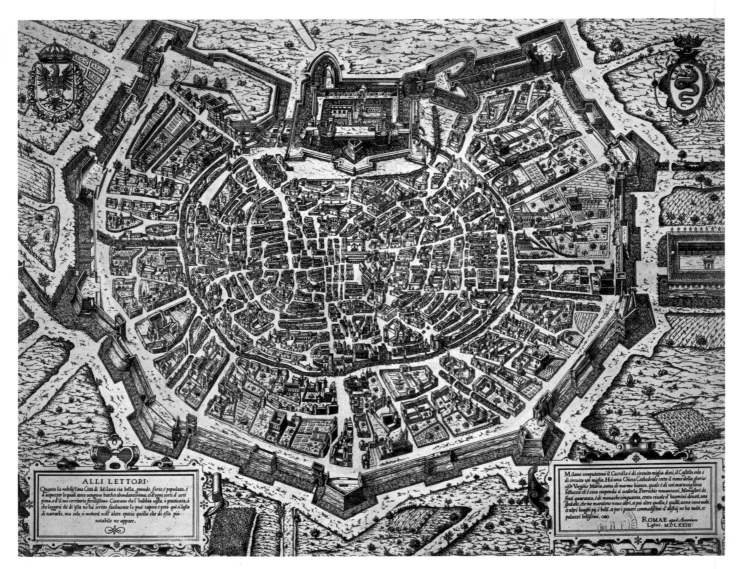

Leonardo—one of the most accomplished lyre players in all Italy—had outplayed all others "gathered together there to play" in a musical competition. Whatever the reason for his trip to Milan, Leonardo kept his eyes and ears open. Lyre music may well have been his ticket to Milan and an introduction to Sforza, but he was not interested in establishing himself in his new city primarily as a musician. Surely aware that Ludovico's chief engineer, Bartolomeo Gadio, was soon to retire, Leonardo's eyes were on the prize of a position as military engineer or architect at court. Further, there was the matter of the not-yet-awarded commission for the great bronze equestrian statue commemorating Ludovico's father. A test of artistic prowess, the work also presented scientific and engineering challenges that would secure the reputation of the man who successfully completed it. Leonardo was surely thinking along these lines when he addressed a famous letter of introduction to Sforza—a letter he may never have sent, but one that reveals his Milan career plan. Though Ludovico may have been enchanted by Leonardo's lyre music, five or six years elapsed before Leonardo's genius was recognized by the Sforza court. In the meantime, Leonardo had to start life in Milan, plying his painting trade as he had in Florence.

In the early 1480s—with Milan's growing wealth and with Ludovico's encouragement—artists in the city were beginning to prosper, and painters, sculptors, architects, and other skilled craftsmen from France, Germany, Flanders, and the rest of Italy were flocking to Milan. The most prominent artist and architect in Milan was Donato Bramante. Born in Urbino in 1444, Bramante had arrived in Milan in the mid-1470s. Like Leonardo, Bramante was an amateur musician—a lute player. Perhaps he and Leonardo met at some music festival or competition; whatever the setting, they became close friends and intellectual companions. Sharing a passion for mathematics and engineering, both wished to move from painting to the purer, more noble profession of architecture, and they desired the patronage of Ludovico in a life at court. Bramante, ahead of Leonardo in realizing his ambitions, secured a prestigious architectural commission, the oratory of Santa Maria. Success followed success, and Bramante later crowned his career as an architect with the design of St. Peter's Basilica in Rome. Leonardo's fortunes, however, lay elsewhere. He never achieved real success as an architect, and he continued to toil as a painter during his early years in Milan. His first employment at court—fortunately for posterity—was as a painter.

"Virgin of the Rocks"

The most prominent studio or bottega in Milan at this time was that of the de Predis brothers. Before the end of 1482, certainly by the first months of 1483, Leonardo became a headline artist in their workshop. To Leonardo the association held out the prospect of contact with important patrons. For the de Predis it offered the cachet of Leonardo's name and its association with the art of Florence, the acknowledged center of Italian painting, sculpture, architecture, and design. In fact, in the contract for the painting known as the "Virgin of the Rocks," Leonardo's first commission in Milan, he is referred to as maestro, or master, while his partners, Ambrogio and Evangelista de Predis, are given no special title. Throughout his time in Milan he was styled as "Leonardo Fiorentino," Leonardo the Florentine, not only designating his earlier place of residence but also distinguishing him from other homegrown artists and sculptors. Yet the name must have been ambiguous to Leonardo; though he was a member of the Florentine tradition of great artists and artisans, this label identified him as an outsider. He had felt the ignominy of being an outsider in Florence—now his new honorific not only designated him as an outsider but also reminded him of that past.

The contract for the "Virgin of the Rocks" is dated April 25, 1483. It calls for the painting of an altarpiece for the chapel of the Confraternity of the Immaculate

"He gave some time to the study of music, and learned to play on the lute, improvising songs most divinely."
—VASARI

LETTER OF INTRODUCTION

"I have plans for bridges, very light and strong and suitable for carrying very easily, with which to pursue and at times defeat the enemy."

—LEONARDO

Among Leonardo's notes is a copy of a letter of introduction to Ludovico Sforza, strongman of Milan. The text that survives in the notebooks is a draft with insertions and revisions, not a fair copy. It is not known whether Leonardo wrote this letter before he left Florence or after he arrived in Milan—and there is no evidence that he ever presented the letter to Sforza. Yet the letter tells us something of Leonardo's ambitions for his life in Milan.

Below: Anonymous, cameo with portrait in profile of Ludovico "Il Moro," late 15th century. Museo degli Argenti, Palazzo Pitti, Florence

Leonardo wanted a position at court, preferably as an engineer, and he must have known that Ludovico's chief engineer, Bartolomeo Gadio, was about to retire. Just as clearly, he understood the needs and interests of his prospective master. The Sforza dynasty in Milan, unlike the Medici one in Florence, was based on military power. Ludovico's grandfather, Muzzo Attendolo, a condottiere (soldier for hire), had adopted the family name Sforza from the Italian *sforzare*, to compel or force; Ludovico's father, Francesco, had overthrown the previous ruling family of Milan, the Visconti. More than most other city-states of Italy, Milan and its rulers lived by the sword.

With this in mind, Leonardo opens his letter with a direct appeal to Ludovico's military needs: "Most Illustrious Lord ... I shall endeavor, without prejudice to anyone else, to reveal my secrets to Your Excellency." Leonardo boasts of the ingenious military inventions in his portfolio—light portable bridges that can be carried by armies, siege engines capable of "destroying any citadel," naval vessels that "resist even the heaviest cannon fire," machines that tunnel under walls without detection, and covered vehicles that "penetrate enemy ranks" and "destroy the most powerful troops." When the letter was written, Leonardo almost certainly had none of these designs on the drawing board. The letter demonstrates Leonardo's brazen confidence in his own abilities. Determined to secure a court appointment, he would worry about the details after he got his foot in the door—and, later

on, he actually made sketches for some of the machinery he mentions, as well as designs for even more marvelous engines of war.

In the closing paragraphs of the letter, Leonardo states that he is "the equal of any man in architecture," a claim as audacious as his claim to be a military engineer. Finally, he mentions some skills for which he could produce a record of accomplishment: "I can carry out sculpture in marble, bronze, and clay; and in painting can do any kind of work as well as any man, whoever he may be." Coming at the end of the letter, this reference to sculpture and painting has seemed to some an afterthought, Leonardo downplaying his stature as an artist. But it is probably a case of Leonardo playing his cards in a careful and calculating way. After his attempt to clinch the deal with all the military marvels that he can bring to Ludovico's arsenal, Leonardo mentions an added bonus—his skills as a sculptor and painter. It is as if he were saying—And by the way, that huge equestrian statue of your father, Francesco, that you envision, I am just the man to do it. In the end he would be.

Opposite: Master of the Pala Sforzesca, "Virgin and Child, the Doctors of the Church and the Family of Ludovico 'Il Moro,'" 1494. Pinacoteca di Brera, Milan.

Above: Leonardo, detail from "Drawing of Soldiers and Armed Chariot," 1485-1490. Codex Atlanticus, Biblioteca Ambrosiana, Milan

> "Our Lady is at the center, her cloak [is to] be of gold brocade and ultramarine blue ... the gown ... gold brocade and crimson lake, in oil ... the lining of the cloak ... gold brocade and green, in oil.... Also, the seraphim done in sgrafitto work."
>
> —CONTRACT FOR THE "VIRGIN OF THE ROCKS"

Conception in the church of San Francesco Grande, the second largest church in Milan after the Duomo. Leonardo and the two de Predis brothers were to produce a triptych. The center panel was to have the Virgin and Child in the midst of a host of angels attended by two prophets. The two side panels were each to contain four angels playing music. Maestro Leonardo was to paint the central panel, Ambrogio the side panels, and Evangelista would repair and gild the already existing frame supplied by the church. The finished work was to be delivered in December 1483, in just nine months, a tight schedule for any such commission—and an impossibly optimistic timetable for any undertaking involving Leonardo.

The contract also specified other details concerning the composition and choice of colors: "Our Lady is at the center, her cloak [is to] be of gold brocade and ultramarine blue ... the gown ... gold brocade and crimson lake, in oil ... the lining of the cloak ... gold brocade and green, in oil.... Also, the seraphim done in sgrafitto work.... Also God the Father [is] to have a cloak of gold brocade and ultramarine blue. The mountains and rocks shall be worked in oil, in a colorful manner." After agreeing to the terms of the contract, Leonardo ignored these presumptuous demands and went on to produce his own version. Even the side panels done by Ambrogio de Predis would not meet the requirements of the contract. They contain

In performing infrared refractography on the "Virgin of the Rocks" (opposite) the technical staff of London's National Gallery discovered that Leonardo had considered a dramatically different composition before settling on the one presently seen. A tracing (left) was made from the infrared study.

one angel each, not four. The central panel would not have the specified host of angels, and there would be no prophets at all, much less cloaks of gold brocade and ultramarine blue.

Instead, the composition comprises the Virgin, the infant Christ, the infant St. John, and an angel, a divine quartet, organized by a pyramidal structure with the Virgin's head at the vertex. The Virgin's right arm is draped over the shoulder of the infant St. John, and her left hand poised over the figure of the infant Christ. In the lower right portion of the panel is the kneeling angel pointing toward the infant St. John. Rugged stalactites and stalagmites, thick columns of rocks rising from the ground—all immersed in a thick mist but punctuated by bright light—provide a surrealistic backdrop.

The painting depicts a scene from a story popular at the time—the meeting of the infant Christ and St. John that took place during the flight of the holy family from Egypt. The story comes from the apocryphal gospel of St James. In presenting a religious narrative of this magnitude, artists were typically expected to invoke symbolic devices. Since Mary, according to the Catholic doctrine of the Immaculate Conception, was supposed to have been conceived unsullied by original sin, Leonardo would paint her flat chested and exuding an ethereal aura of innocence and piety. Leonardo also includes symbolic flora, depicting them just as he saw them in nature instead of copying lithographic images provided by herbalists, as was the practice of most other artists. As suggested by Maxine Anabell, the choice of flora included "Aquilegia, or columbine (dove plants) beside the Virgin's face; these symbolise the Holy Spirit. Stains on the St. John's wort suggest a martyr's blood, the creeper Cymbalaria symbolises constancy and virtue. Heart-shaped leaves represent love and virtue; sword-shaped leaves, the sword of sorrow that was to pierce Mary's heart, and the palm leaves are a symbol of victory." The four characters and the symbolic flora are placed against an unforgiving rocky background—a harsh reality framing the soft images of the plants. Leonardo, obsessed with nature and possessing deeper understanding of both geology and botany than any peer—artist or scientist—rendered the diametrically opposing contrasts of the soft and hard, light and dark, pure and sullied more effectively than any other painter had ever done.

Two versions of the "Virgin of the Rocks" exist, created about 15 years apart. The first is considered technically superior, although some scholars claim the second version reveals a more mature Leonardo. Hanging in the Louvre, not far from the "Mona Lisa," the first is thought to be wholly by Leonardo. The second version, hanging in London's National Gallery, was probably painted by an assistant—perhaps Ambrogio de Predis—but overseen by Leonardo. The London version is now flanked by Ambrogio's panels originally painted for the Louvre version.

A great deal of uncertainty and speculation surrounds the historical ownership of both works, but a reasonable guess is that the painting hanging in the Louvre found

Opposite: Leonardo, "Virgin of the Rocks," ca 1483. Oil on wood, transferred to canvas in 1806. Musée du Louvre, Paris

Below: Raphael, "Alba Madonna," 1510. Oil on wood, transferred to canvas. National Gallery of Art, Washington, D.C.

Following pages: Bird's-eye view of people and pigeons in the Piazza del Duomo, Milan, from the roof of the Duomo

PYRAMIDAL COMPOSITION

Leonardo introduced pyramidal construction in his early, unfinished "Adoration of the Magi"; he employed it again in both the Louvre and London versions of the "Virgin of the Rocks." Significantly, this use of pyramidal construction in his painting was to have a very long reach. Several of Raphael's paintings, including the "Alba Madonna" (National Gallery of Art, Washington, D.C.) and "Madonna of the Chair" (Palazzo Pitti, Florence), display pyramidal composition inspired by Leonardo's paintings. Even Michelangelo, who outwardly expressed more disdain than admiration for Leonardo, was clearly influenced by his rival, employing the pyramidal structure in his painting the "Holy Family and the Infant St. John" (1504), also known as the "Doni Tondo," now hanging in the Galleria degli Uffizi, Florence.

an eventual home there after passing through the hands of French royalty. The painting was at the center of a series of lawsuits and petitions beginning in about 1492. There is speculation that it may have been bestowed as a token of gratitude by Leonardo to King Louis XII, who had ruled in his favor in the lawsuits in 1506. But this theory casts the king in an unfavorable light, settling a lawsuit in a manner that quickly redounds to his benefit—not exactly building his reputation as an impartial judge. Further, this theory does not explain how the second version came into existence, or how the second version became associated with the side panels of the original version.

Another theory, supported by Charles Nicholl, author of a recent biography of Leonardo, speculates that Ludovico Sforza was behind the initial lawsuit of Leonardo and the de Predis brothers to regain possession of the painting from the Confraternity on the grounds of inadequate compensation. The suit claimed

"*A good painter has two main objects to paint, man and the intention of his soul. The former is easy, the latter hard as he has to represent it by the attitude and movement of the limbs.*"

—LEONARDO

Leonardo, "Portrait of Cecilia Gallerani," aka "Lady with an Ermine," late 1480s. Czartoryski Museum, Krakow

a willing purchaser existed—ready to pay what the petitioners were seeking and the Confraternity of the Immaculate Conception was unwilling to pay. According to this theory, the fate of the first version was settled when the artists agreed, after repossessing the painting and selling it to the willing purchaser, Sforza, to supply the Confraternity with a replacement copy, the second version. Ludovico then presumably presented the first version to the Emperor Maximilian as a present on the occasion of the emperor's marriage to Ludovico's niece, Bianca Maria, in 1493, shortly after the initiation of the lawsuit. This reconstruction of events is supported by Antonio Billi, the earliest of the contemporary biographers of Leonardo. The "Virgin of the Rocks" is the only altarpiece Leonardo painted while in Milan. In this version of the story, the painting would have found its way to France when Maximilian's granddaughter Eleonora married François I in about 1528. By 1625 the painting was hanging in the Fontainebleau Palace in France, along with the "Mona Lisa," before finding its way into the Louvre.

This story of how the two versions came into existence also explains why the London version is surrounded by the original panels painted by Ambrogio de Predis. They were not sent as part of Ludovico's wedding present. The London version is the copy produced for the church of San Francesco Grande when the Confraternity specified a facsimile of the earlier work. The painting has overall proportions very close to its predecessor, but has a somewhat cooler, bluish tinge. The backdrop of stalactites and stalagmites has been brought forward slightly; the Virgin and the two infants have halos, and St. John holds a cross of reeds; and the angel is not pointing in the direction of St. John.

Before the end of the decade, Leonardo was working out of his own studio in Milan. Ambrogio de Predis probably joined Leonardo in his new bottega. The second version of the "Virgin of the Rocks" was likely a product of this studio and Ambrogio de Predis's brush. Marco d'Oggiono and Giovanni Boltraffio, mentioned specifically by Vasari, were apprentices and assistants in this first Milan studio, and were responsible for many Leonardesque productions such as Boltraffio's "Madonna and Child." Though touched by the master's hand, the "Madonna Litta" is most likely the work of Marco d'Oggiono. Paintings of the Madonna and Child were a favorite product of the studios of the late 1400s, and Leonardo's workshop turned out its share.

Cecilia—the Lady with an Ermine

For Leonardo, recognition at the court of Ludovico Sforza did not come until the end of the 1480s. The "Virgin of the Rocks" had greatly enhanced his reputation as a painter, for almost immediately after its completion it was widely regarded as an innovative and noteworthy piece of work. And, as mentioned, it may have attracted the attention of Ludovico. It is likely that not long after the establishment of his own studio Leonardo landed his first official job for Ludovico in the form of a commission for a painting. He was hired to do a portrait of Il Moro's mistress, Cecilia Gallerani. The two most important paintings produced by this new studio, wholly or mostly by the hand of Leonardo, were the portrait of Cecilia Gallerani—one of his masterpieces—and the portrait of a young musician.

Cecilia was born in 1473 to Fazio Gallerani and Marguerite Busti. Fazio had served in the Milanese diplomatic service as an ambassador; Marguerite was the daughter of a successful lawyer. With this parentage, it is easy to understand why Cecilia enjoyed a reputation as a cultivated and beautiful woman. Probably sometime in the late 1480s, when she was 14 or 15, she became the mistress of Ludovico, despite his being betrothed to Beatrice d'Este, daughter of the Duke of Ferrara. Cecilia herself was under childhood marriage contract to Giovanni Stefano Visconti, until released from this obligation in 1487—perhaps because of her liaison with Ludovico. In late 1490 she was pregnant with Ludovico's child. In January 1491 Ludovico married Beatrice as planned, and in May Cecilia gave birth to a son, Cesare Sforza Visconti. Just before the birth, Cecilia moved out of the palace, ending the liaison with Ludovico. She took her famous portrait with her.

In the painting, the subject appears older than the teen she would have been in the late 1480s when Leonardo created the portrait; during the Renaissance very young women often contrived to look older than their years, and an artist would

> "The mind of the painter should be like a mirror which always takes the color of the thing that it reflects, and which is filled by as many images as there are things placed before it."
>
> —LEONARDO

certainly have cooperated in the deception. There is the possibility that the painting depicts a slightly older Cecilia and was painted as late as 1491, but not later than 1492, for the poet Bernardo Bellincioni, who died in 1492, wrote a sonnet about the portrait, using the description, "She seemed to listen and not to speak."

The portrait depicts the subject gazing soulfully into the distance; Leonardo used the artistic device of composing the portrait in a gentle helical twist—a technique he employed to deliver dynamism to her posture. Her left arm and the slender fingers of her right hand cradle a ferret, long miscast by art historians as an ermine. (The two animals strongly resemble each other, but the ermine has never been domesticated, whereas the ferret has been domesticated for several thousand years.) Ferret or ermine, Leonardo must have thought it was an ermine, a symbol for chastity. Fond of invoking plays on words, Leonardo would have enjoyed the subtle joke of using an animal whose name in Greek, *galee*, evokes the subject's name, much as he had done in using the juniper bush, *ginepro* in Italian, as the background in the portrait of Ginevra. Also, in 1488 Ludovico had been given the title "L'Ermellino" by the King of Naples, and had been styled, again in a poem by Bellincioni, as *bianco ermellino,* the white ermine. Cecilia cradles the "ermine" in her arms in the painting, but it is powerful and muscled, and has the air about it of a wild animal.

Studying the details of Cecilia's dress, one can make out the elegant velvet and lace, all in dark and subdued colors so as not to detract from the prominence of the brightly lit face. Unlike the backgrounds of the rolling countryside in the earlier "Portrait of Ginevra de' Benci" and the later "Mona Lisa," the background in the "Cecilia" is nearly monochromatic and heavily darkened, except for speckles of turquoise showing through where the dark background has flaked off. But modern technology reveals that the background has been heavily overpainted (perhaps as late as 1800) by another artist and that the original background consisted of gradations of blue-green rather than a landscape. Finally, in the top left corner of the darkened background there appears the unevenly printed title in capital letters, "Belle Feroniere Leonard Dawinci," associating the painting with its Polish owners, the Czartoryski family.

Cecilia visited Washington, D.C., in 1992 in order to participate in the celebration of the 500th anniversary of Columbus's expedition to the New World. The agreement between the Czartoryski Museum and the National Gallery of Art called for the exhibition of the painting for three months, followed by its examination by the specialists of the National Gallery's conservation laboratory for an additional week. David Bull, who in his professional career had cleaned and restored masterworks by great artists, including Bellini, Titian, Raphael, and Rembrandt, had earlier discovered Leonardo's fingerprints in the bushes in the National Gallery's "Portrait of Ginevra de' Benci." He recalled the thrill of having Leonardo's two girls not just in the same building but "on the Formica table, side-by-side…. In the rarified world of art masterpieces such a moment is without compare." Again he found Leonardo's fingerprints, this time on the neck of the portrait of Cecilia, evidence of Leonardo's use of the velature technique—of applying colors directly onto the panel, and using his fingertips to blend them. But lacking the optimum equipment to photograph the fingerprints, the gallery staff invited the FBI to photograph them with their specialized cameras. Leonardo's fingerprints are now in the archives of the Bureau.

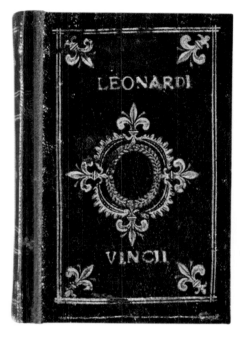

Above: Front cover of Leonardo's Manuscript K, 1503–08. Institut de France, Paris

Opposite: Leonardo, "Portrait of a Musician," ca 1490. Pinacoteca Ambrosiana, Milan. It is possibly a portrait of Franchino Gaffurio.

"Portrait of a Musician"

The so-called "Portrait of a Musician," a painting probably of the late 1480s, is believed to be mostly by the hand of Leonardo. There is no certainty about who is depicted in the painting. At one time it was thought that the portrait might even be of Ludovico Sforza. A more probable suggestion is Franchino Gaffurio, the *maestro di cappella* of the Duomo of Milan. Painted on the same species of wood as the portrait of Cecilia, the young subject is shown in three-quarter profile,

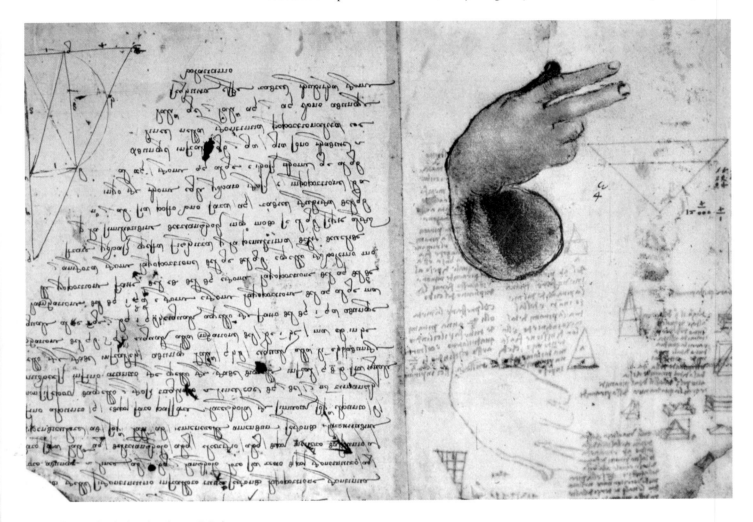

Leonardo, "Study of a hand and notes." Codex Atlanticus, Biblioteca Ambrosiana, Milan

gazing into the distance, and clutching a sheet of paper in his hands. He displays a distinctly noble countenance, soulful eyes, a strong jaw. The understanding of the bone structure of the subject's face, and especially the spiraling curls, are unmistakably Leonardesque. The overall composition and the face are most certainly by Leonardo, but the rest is believed to have been worked on by one of his assistants, and then overpainted in some areas by a 19th-century restorer.

The hat and vest the subject is wearing are by a painter of distinctly lesser skills than those of Leonardo. The areas painted by other hands display right-handed brushstrokes. Although Leonardo wrote and sketched almost exclusively with his left hand, when applying paint he demonstrated ambidexterity.

Moreover, he applied the colors mostly with his fingers. Thus it is not the direction of the brushstrokes but the quality of their application that is the clue to the work by lesser talents.

For three centuries the sheet in the subject's hand appeared illegible; then a thorough cleaning and restoration effort in 1905 uncovered a musical score, which gave the painting its modern appellation, "Portrait of a Musician." In view of Leonardo's fondness for mental exercises, the musical composition has attracted the attention of cryptologists and musicians trying to uncover hidden messages in the score—so far to no avail.

The Notebooks

In the mid-1480s, probably around the time of the completion, or shortly after the completion, of the original "Virgin of the Rocks," Leonardo began the regular practice of keeping a notebook. There are earlier sketches and jottings on single folio sheets from his days in Florence that were later bound into miscellaneous collections. But they were intermittent and haphazard, not the sustained record of intellectual inquiry found in the notebooks. There is no evidence, until the mid-1480s, that Leonardo was steadily engaged with the science and technology he pursued with relentless enthusiasm from that time forward.

The notebooks are his intellectual autobiography, the dwelling place of the Leonardo who fascinates us today. Here we encounter the working scientist-engineer. Out of an original hoard of material—which numbered perhaps 15,000 to 20,000 pages at his death, a lifetime of pondering, questioning, inventing, and sketching—perhaps a quarter survive. These materials were bequeathed to a favorite apprentice, Francesco Melzi, who kept them largely intact and in his possession until his own death in 1579. Melzi's heirs were less concerned with keeping the great man's legacy together. Parts of the collection were sold; other parts were given away, sometimes just a few pages at a time. This disorganized disbursement led to the eventual loss of much of the material. The existing pages survive in three forms: bound miscellanies put together after Leonardo's death, notebooks that remain largely as they were when Leonardo was alive, and single sheets.

Two of the most famous and important of the surviving collections are the Codex Atlanticus in the Biblioteca Ambrosiana in Milan, originally (until reorganized and rebound in the 1960s) a single volume of some 481 sheets containing some of the master's most famous drawings, and the collection of the Royal Library at Windsor Castle in England, consisting of about 600 unbound folio-size pages of drawings and writings, including some of the most impressive of Leonardo's anatomical sketches. The Institut de France in Paris holds the largest collection of small notebooks in the same arrangement and condition (and in some cases the same binding) as they were left by Leonardo. These were appropriated from the Biblioteca Ambrosiana by Napoleon while on campaign in Italy in the 1790s. Other collections exist in Milan, Turin, London, Seattle (in the collection of Bill Gates), and Madrid. The Codex Madrid comprises two complete notebooks that were discovered in the National Library of Spain in 1967, after being misfiled for centuries.

The pages of the manuscripts and notebooks reveal the working methods of Leonardo the artist as well as the scientist. Some of the shorter notebooks are devoted

LEONARDO'S MIRROR TEXT

In recording his observations, Leonardo employed a "mirror text," written backwards, from right to left. Biographers have often assumed that he was being secretive. But the more likely explanation for this particular quirk is simpler. The shading in all his sketches—produced by strokes from lower right to upper left—reveals that he was left-handed. Right-handers characteristically shade with strokes from lower left to upper right. Pushing a pen (in his time, a quill) to the right by a left-hander can cause the point to dig into the paper. Moreover, by writing from left to right in the conventional manner, he would have smeared the fresh ink with his hand. Leonardo's observations and speculations were carried out mainly to satisfy his own curiosity and were written as a personal record, rather than for perusal by others. He did not bother to observe conventions while capturing the output of his teeming mind, which never dwelt long on any particular project or idea before being taken away by the next.

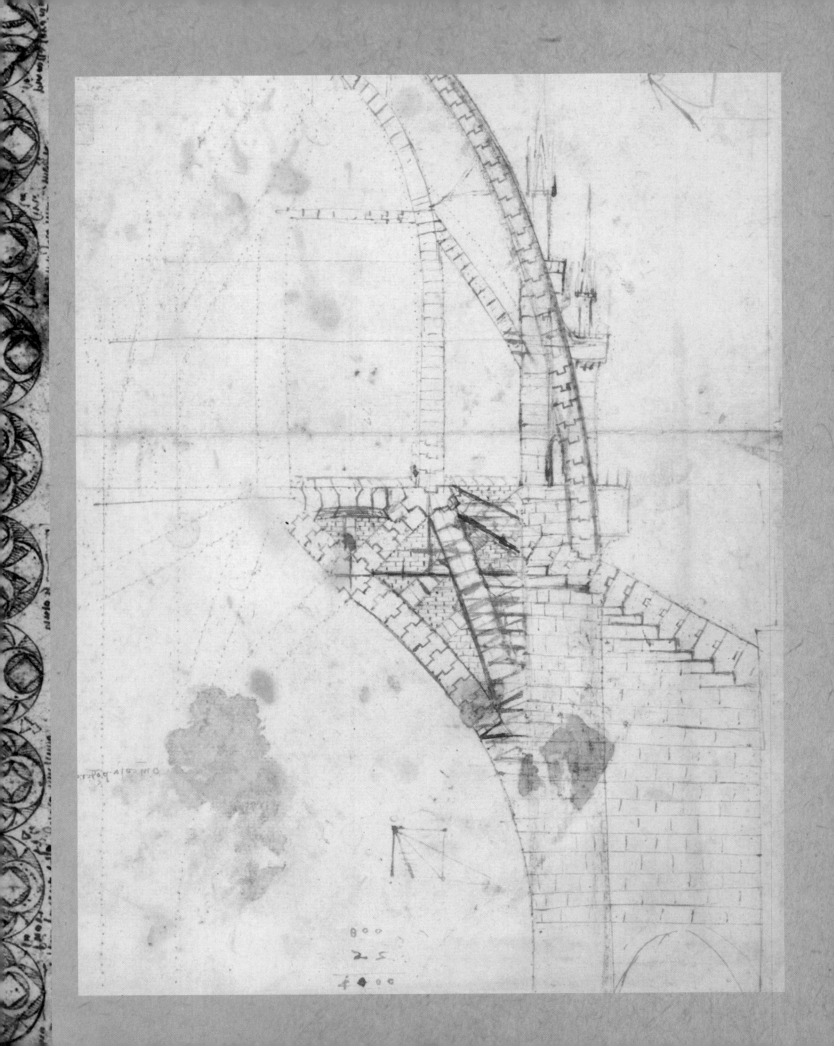

LEONARDO AS URBAN PLANNER

"Practice must always be founded on sound theory, and to this perspective is the guide and the gateway. And without this nothing can be done well in the matter of drawing." —LEONARDO

Opposite: Leonardo, "Drawing for the Dome of the Duomo of Milan," ca 1487. Codex Atlanticus, Biblioteca Ambrosiana, Milan

Above: Leonardo, "Architectural Study for a City on Several Levels," ca 1487. Paris Ms B, Institut de France, Paris

In 1485, while Leonardo was probably at work on the "Virgin of the Rocks," the bubonic plague had descended upon Milan. The epidemic was devastating, wiping out as much as a quarter of the city's population. For Leonardo, who took great care with sanitation and personal hygiene and found the foul smells of city life offensive, the outbreak must have been not only horrifying but also thought provoking. The fastidious Leonardo—in contrast to his rival Michelangelo, who rarely bathed or washed and often wore grimy clothes—no doubt found the close crowds, confined spaces, and dirty streets of Milan a cause for concern, as well as an inspiration for better city planning.

Though not especially known for his architectural accomplishments, Leonardo did submit, circa 1487, a proposal for completion of the dome of the cathedral in Milan. It was a double-skinned dome, similar to Brunelleschi's famous dome in Florence, which he knew intimately. Leonardo's design remained in the running until the very end, but the authorities ended up choosing a less daring plan. Leonardo responded with an even more daring proposal for city redevelopment. No doubt with the recent plague in mind, he conceived a double-tiered city. The upper residential deck has open piazzas, airy second-story loggias, and broad avenues for pedestrians. Houses are no higher than the streets are wide, allowing sunlight to penetrate to the sidewalks. Chimneys are tall enough to carry smoke away from the eyes and lungs of the residents. Staircases are circular in order to eliminate the corners often used as public urinals. Animal traffic, warehouses, and the trades are confined to the lower level, where drains carry off waste and sewage, and canals are used for the transport of goods. The plan was never considered by Milan or any other city, but it is one of the first times in the notebooks that we see Leonardo the Futurist, thinking, inventing, and pushing the boundaries of the possible.

almost exclusively to a single subject, such as the flight of birds or the use of light and shade in painting, but most are a jumble of material reflecting both his wide-ranging interests and his habit of jumping from one line of inquiry or invention to another. They are encyclopedic in their embrace. The can range over anatomy, mathematics, architecture, astronomy, military engineering, aerodynamics, botany, optics, music, studies for paintings—all in no particular order, often in the space of a relatively small number of consecutive pages. Leonardo, in quest of universal knowledge, is determined to attain his goal through his own observations, speculations, and experiments, rarely accepting existing authority on any subject.

"Science is the observation of things possible, whether present or past. Prescience is the knowledge of things which may come to pass, though but slowly."
—LEONARDO

We can only speculate about what prompted Leonardo to begin the regular practice of keeping a notebook. Was he simply at wit's end in Milan—with time on his hands, not knowing where his next commission would come from, and frustrated in his ambitions in this new home? Did serious hopes of securing a post at court as a military engineer require him to produce a portfolio of convincing designs of military equipment? Or was he conscious of his own mortality because of the plague then raging through the city that would take a quarter of the city's population? The oldest surviving notebook, usually designated as Paris Ms B, in the Institut de France, contains notes and plans for an ideal city, in which concern for sanitation was a prominent feature. Though the true nature of contagious disease was little understood in the 15th century, Leonardo seems to sense that airiness, sunlight, and cleanliness could contribute to better health.

Thoughts on Architecture

During the late 1480s, Leonardo filled his notebooks with exterior drawings and floor plans for a series of buildings organized around a central dome. These buildings resemble nothing so much as a cathedral though contemporary scholars do not believe a cathedral was intended. Whatever the purpose of these buildings, we can

Opposite: An elevated view of Milan's Duomo reveals the cathedral's intricate architecture. Begun in 1386, and completed in the 19th century, it is the second largest church in Italy after St. Peter's Basilica in Rome.

Above: Leonardo, "Study for a Building on a Centralized Plan," ca 1487-1490. Codex Ashburnham, Institut de France, Paris

see the influence of the Duomo in Florence and Brunelleschi's ribbed dome. Leonardo is competent as an architect, but largely derivative. His architectural vision is certainly futuristic, but not unusual for his time. Making designs and specifications for an ideal city was a favorite Renaissance pursuit. Other than his ideas for improved sanitation, he offers little that is original or innovative, although his ideas are often practical. Later, when he is finally employed as a military engineer, he will design fortifications that take into account the increased presence of artillery in warfare, displaying sloped walls with a broadened base to prevent undermining by continuous bombardment, but others were already on the same track. Few of his designs for buildings astound us as do so many of his drawings of machinery and inventions. Most of the architectural studies in these early notebooks before 1490 are rough sketches, not executed with the meticulous care and attention to detail that will be so evident in later anatomy and engineering studies. But he is making a beginning, learning the craft of draftsmanship and mechanical drawing, which he will eventually take into uncharted territory.

Leonardo did have one serious encounter with real architecture during these years. In 1487 there was a competition for the design of a new dome for Milan's cathedral. Leonardo, as well as his close friend architect Donato Bramante, submitted proposals. Leonardo even received payment for the construction of wooden models of his design, which was a double-skinned dome similar to Brunelleschi's design for the Duomo in Florence. There is even a draft of a speech in one of his notebooks, apparently intended as part of a presentation to be given before whatever committee of experts or elders would make the final selection of the winning proposal. Delay followed delay on the part of the cathedral authorities, and Leonardo's interest in the project waned before the final selection of a design by two architects of Lombardy, Amadeo and Dolcebuono, in June 1490.

The early notebooks are filled with Leonardo's first sustained attempts at military engineering. Like the architectural drawings, their sudden prominence in the notebooks reflects Leonardo's abiding ambition for permanent employment at court. Leonardo will turn out to be about as good at military engineering as he was at architecture—competent though not brilliant. Just as none of his sketched buildings were ever built in his lifetime, none of his military machines were ever constructed. But among his designs for engines of war, a few are truly pioneering and forward-looking, though impractical in his time. Some of the military hardware even looks backward. Leonardo filled pages with designs for pikes and halberds, and he is fascinated by variations on the catapult, trebuchet, and crossbow, though these were becoming anachronisms at the end of the 15th century.

Firearms, artillery, and explosives were the new face of war, and this was not lost on Leonardo. His most imaginative inventions anticipate the future. He designed a multibarreled gun that would have been the first machine-gun, a submarine, and an armored car sometimes referred to as the first tank. It appears on the same sheet as his scythed chariot, a frightening contraption meant to mow down attackers during a siege. He reinvents Archimedes' "architronito," or steam cannon, where water is poured into the heated cannon and the steam produced under pressure launches the projectiles.

Though never leading to the actual production of equipment, these military drawings were a valuable exercise for Leonardo. Even the unorthodox catapults and machines like the repeating crossbow, though looking to the past, were for Leonardo the start of his encounter with complex devices and their relation to the forces of nature. In rendering these engines, Leonardo further refined his mechanical drawing until it was ready for the careful depiction of the complicated geared machinery he invented in the 1490s, machinery that 350 years later powered the industrial revolution.

Leonardo's Dream

In these early notebooks we also find Leonardo engaged with his lifelong interest in flight—the flight of birds and the possibilities of human flight Horizontal and

Leonardo, "Sketch of Birds in Flight," 1490-1505. Codex Atlanticus, Biblioteca Ambrosiana, Milan

vertical versions of his now famous ornithopters—machines with birdlike flapping wings powered by a human passenger—are found in his very first notebook. These machines were not practical, and, in fact, would not have worked: The power necessary to drive them could not be generated by the single man for whom they were designed. But they are examples of magnificent failures; first ideas, while not successful, still tell us what changes the next version should incorporate or what new questions should be asked and answered. Leonardo never realized his dream of human flight, but he learned a lot about aerodynamics along the way. The ornithopters are a testament to the reach of his imagination. And his unwillingness to abandon the study of flight—or the possibility of human flight, despite lack of any positive result—is an expression of his belief that nature, even if she doesn't always

IDEAL PROPORTIONS

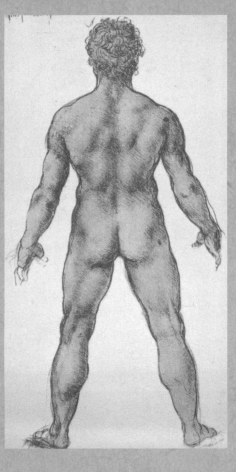

> "Man is the measure of all things.... Every part of the whole must be in proportion to the whole."
>
> —LEONARDO

Leonardo, in the manner of the pure scientist, made precise measurements in nature, sought connections and patterns, and expressed them mathematically. He counted branches of trees, veins in leaves, petals in flowers. He measured parts of the face, and the lengths of the limbs. And as a scientist-artist, he felt compelled to imbue his art with the precision of the results springing from his scientific inquiry. He wrote about ideal proportions in the face: "The space between the slit of the mouth and the base of the nose is one-seventh of the face.... The space from the mouth to below the chin will be a quarter part of the face, and similar to the width of the mouth.... The space between the chin and below the base of the nose will be a third part of the face, and similar to the nose and the forehead. The space between the midpoint of the nose and below the chin will be half the face."

In a number of his drawings, geometric constructions are superimposed on faces of humans and even on horses. There are golden rectangles, golden and equilateral triangles, and more. These all represent his interest in achieving ideal proportions.

VITRUVIAN MAN

In the first century B.C., the Roman architect Vitruvius proposed in his book, *De architectura,* that the proportions found in the human body were ideal for designing buildings. Vitruvius wrote: "The navel is naturally placed in the center of the human body, and, if in a man lying with his face upward, and his hands and feet extended, from his navel as the center, a circle be described, it will touch his fingers and toes. It is not alone by a circle, that the human body is thus circumscribed, as may be seen by placing it within a square." Artist and theoretician Leon Battista Alberti trumpeted the same proposition in his 1435 treatise, *On Painting,* a book known to have been among Leonardo's cherished collection. Artist Cesare Cesarino was one of Leonardo's many contemporaries who attempted to demonstrate the Vitruvian conjecture. In his book *Di Lucio Vitruvio de Architectura* (ca 1521), Cesarino presented his Vitruvian man—limbs stretched out corner to corner in a square, the square circumscribed by a circle—with the man's naval coinciding with the center of the circle, and with the diagonal center of the square. His scheme, as in the attempts by other artists, makes the length of the limbs disproportionate to the size of the torso—the arms are too long, the legs too short. The man appears about to be drawn and quartered.

Leonardo sought to achieve correct perspective and proportion, but also to infuse the work with the spiritual symbolism prevailing at the time. The square was to represent the earth and the mundane, and the circle heaven and the spiritual. For Leonardo, the Vitruvian man, representing the mechanism of the human body, was emblematic of the mechanism of the universe itself, a deep-seated belief in the *cosmografia del minor mondo* (cosmography of the microcosm). Leonardo created his Vitruvian man in an ingenious and fundamentally different manner than had other artists: The centers of the two geometric figures do not coincide. When the man stands erect with his feet together, the height of the square is defined by the man's height, and the width by his arm-span. The center of the square is located in the man's pubic area. Next the man is depicted with his arms raised, so that the fingertips of his outstretched arms are seen at the level of the top of his head; and his legs spread, so that the span between his feet and the legs themselves create the three sides of an equilateral triangle. In this configuration the man's feet and the tips of his outstretched hands are all right on the circle (close to, but not exactly at the vertices of the square). With the precision of the scientist, Leonardo specified, "If you open the legs so as to reduce the stature by one-fourteenth and open and raise your arms so that your middle fingers touch the line through the top of the head, know that the center of the extremities of the outspread limbs will be the umbilicus [navel], and the space between the legs will make an equilateral triangle."

This may have come at a time when Leonardo was also preoccupied with solving the ancient mathematical problem of "squaring the circle"—of constructing a square with the same area as a given circle in a finite number of steps using only a compass and straightedge. The impossibility of finding a solution was not demonstrated until 400 years later, in 1882.

Above: Leonardo, "Drawing of a Nude Seen from Behind," ca 1503-07. Royal Collection, Windsor Castle, U.K.

Opposite: Leonardo, "Vitruvian Man," ca 1490. Galleria dell'Accademia, Venice

yield easily, will eventually give up her secrets to the curious, persistent questioner. He had, in other words, something nearly unique in his time: a modern scientific temperament that saw failure to be as great an opportunity as success.

His notebooks from 1485 on abound with sketches of bird wings, bird anatomies, and their precise mechanical motion. He regarded birds as nature's splendid flying machines, where form had followed function magnificently. He reasoned that if birds—creatures clearly heavier than air—could fly, then with some help, so could people. The clues, he reasoned, rested in the proper understanding of nature and natural laws. His designs included not only his ornithopters, but also an aerial screw, similar in principle to the modern helicopter, as well as wings for gliding. In one drawing he shows how to determine experimentally the center of gravity of a bird, the sort of understanding that modern aeronautical engineers require in their design of airplanes. In another he sketches out a machine to test the downward force, or lift, created by a fan-like wing connected to a hinged lever.

In order to understand the dynamics of airflow, Leonardo studied the patterns in the turbulent flow of water. With extraordinary prescience, he realized that all fluids behave in a similar manner—for example, vortices in any fluid are caused by analogous perturbations. This is an effect not so readily seen in air but visible in water. Leonardo's understanding of fluid dynamics in water allowed him to make the connections between vortices in water and those in air, and thus to explain the origin of dust devils, downdrafts, or shear forces that can cause airplanes to experience a catastrophic drop. His drawings show an understanding of the existence of vortices within vortices, in precisely the same manner as in modern aerodynamics. No wonder some of his drawings are frequently found in textbooks of aeronautical engineering—and that Leonardo is seen as a founding father by practitioners in that field, as in myriad others. Regarding the turbulence in air, he wrote:

> It often happens that, when one wind meets another at an obtuse angle [an angle between 90° and 180°], these two winds circle around together and twine themselves into the shape of a huge column; and becoming thus condensed, the air acquires weight. I once saw such winds, raging around together, produce a hollow in the sand of the seashore as deep as the height of a man, removing from it stones of considerable size, and carrying sand and seaweed though the air for the space of a mile and dropping them in the water, whirling them around and transforming them into a dense column, which formed dark thick clouds at its upper extremity.

In this early example of his practice of cobbling together diverse fields, he is doing fundamental science, mechanics, and fluid dynamics; and he is prefiguring the fields of hydrology, hydrodynamics, and aerodynamics hundreds of years before they were formally invented.

Leonardo's drawings of fluttering wings in dynamic serial motion—more than a hundred in all, scattered among 13 different notebooks—suggest that he may have possessed preternatural vision, the ability to freeze motion in time, in the manner of the great baseball hitter Ted Williams, who claimed he could see the seams of a pitched baseball approaching him, typically at speeds of 90 miles an

"Often passing by the places where they sold birds he would take them out of the cages, and paying the price that was asked for them, would let them fly away into the air, restoring to them their lost liberty."

—VASARI

hour. Leonardo's drawings reveal that he understood that the flapping is not simply an up-and-down action, but a complex wave motion with a retrograde component; that lift is created not by cupping and pushing down the air, but rather by utilizing Bernouilli's principle—a net upward force resulting from the difference in air flow between the upper and lower surfaces of the wing.

Leonardo also considered achieving elevation with an alternate scheme, using a spinning helical foil. He must have connected that shape and the memories of watching samaras—winged nuts or achenes containing a single seed—descend gently from maple and ash trees while spinning. Could the helical screw be rotated

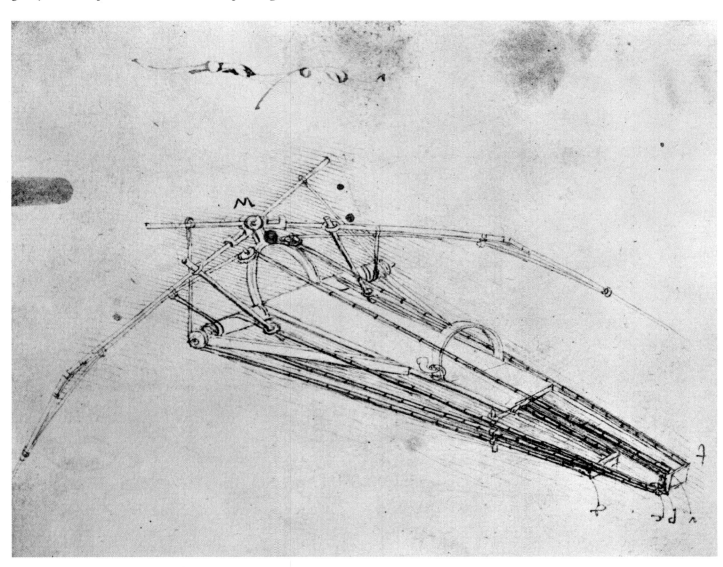

to achieve lift? Leonardo's design was a helicopter with a large helical foil powered by two (and in another version by four) men. A horizontal torque generated by cranking handles was converted into a vertical torque with one of the gear assemblies sketched out in his notebooks. Powered by humans, the aerial screw could never have generated the "bite" necessary to achieve liftoff; moreover, it needed a stabilizer, in order to avoid the spinning of the body of the flying machine, and a steering mechanism to make it practical. His human-powered ornithopters similarly lacked the power to

SLIPPING THE BONDS OF EARTH

"Why should man not be able to do what the birds do?"

—LEONARDO

Five hundred years ago Leonardo expressed an aspiration to achieve manned flight "in the manner of birds." Three hundred years later in France humans would get a first taste of a bird's-eye view, when lighter-than-air flight was achieved. It started with experiments made by the Montgolfier brothers, Joseph Michel (1740-1810) and Jacques Étienne (1745-1799). When air is heated, it expands, and in the process, its density falls below the density of cool air. Accordingly it rises. From conception to implementation, the hot air balloon came to fruition within two years (1782-1783). Two of the Montgolfiers' experiments were especially memorable: On September 19, 1783, a balloon, carrying a number of farm animals (a sheep, a duck, and a rooster), rose from the grounds of the Palace of Versailles to an elevation of 1.6 miles. Then just two months later, on November 21, 1783, the first manned flight of a hot air balloon took place, when the Marquis d'Arlandes and Jean Pilâtre de Rozier, a physicist, flew over Paris, becoming the first test pilots as well as the first air passengers. They achieved a height of 3,000 feet and traveled approximately nine miles during a 23-minute flight.

HEAVIER-THAN-AIR FLIGHT

Four hundred years after Leonardo's prescient musings, the Wright brothers fulfilled his dream, constructing essentially a "manned kite" powered by a gasoline engine; they managed to stay aloft for 59 seconds and traveled 984 feet. This first ever heavier-than-air flight took place in Kitty Hawk, North Carolina, on December 17, 1903. In 1939, Russian-American aeronautical

Opposite: Leonardo, "Study of Mechanics: Aerial Screw or Helicopter," 1489. Brown ink drawing. Paris Ms B, Institut de France, Paris

Above: Bryan Allen flies the pedal-powered Gossamer Condor, designed by Dr. Paul MacCready.

engineer Ivan Sikorsky produced a rotary-wing flying machine—the first successful helicopter—powered by a gasoline engine and stabilized with a propeller in the rear. The inspiration for the design of the helicopter, Sikorsky claimed, harked back to 1901, when at 12 years of age, he had first seen Leonardo's rough sketch of the aerial screw. The dizzyingly swift progress of flight technology culminated just 66 years after Kitty Hawk (July 20, 1969), when three NASA astronauts, launched from Florida on a multi-stage Apollo rocket, reached the moon. One of them, Michael Collins, stayed in a lunar orbit, while the other two—Neil Armstrong and Buzz Aldrin—descended and walked on the moon; all three then returned safely to Earth.

Two nagging questions lurk in Leonardo's notes on flight: Is human-powered flight feasible? Will heavier-than-air flight by a mechanical flapping-wing device work? Inventors working from Leonardo's sketches and probing in the spirit of Leonardo have resolved these questions.

HUMAN-POWERED FLIGHT

In 1977 aeronautical engineer Paul MacCready created the Gossamer Condor, the first human-powered aircraft to fly and maneuver. The plane, with a wingspan of 90 feet and weighing only 70 pounds, took off and flew in a figure-eight pattern around a pair of pylons. Just two years later, Dr. MacCready unveiled the Gossamer Albatross, which crossed the 22-mile-wide English Channel.

POWERED ORNITHOPTER

A functioning flapping aircraft—a powered ornithopter that flew in the manner of a bird, mimicking Leonardo's vision—was constructed as a one-fourth-size scale model, and flown remotely in 1991 by Canadian aerospace engineering professor James De Laurier. It took another 15 years for De Laurier to scale his machine up to full size. On July 8, 2006, flapping at roughly 1.3 times per second and describing the proper wave motion first conceived by Leonardo, Jack Sanderson piloted the gasoline-engine flying machine.

Others dreamed of flying, but Leonardo was the only man of his time to carry out systematic experiments, produce a panoply of designs, and ask just the right questions. We know that on January 3, 1496, Leonardo tested one of his human-powered flying machines unsuccessfully; however, a puzzling entry is found in his notes, its date unknown, one that suggests he might yet have succeeded: "When once you have tasted flight, you will forever walk the earth with your eyes turned skyward, for there you have been, and there you will always long to return."

lift off the ground. But in his experiments immediate success was not as important as his vision of the possibilities—Leonardo was never afraid to experiment and always saw failure as an opportunity to learn.

In one case, Leonardo did not fail, although the design was only demonstrated to be practical nearly 500 years after its conception. Leonardo's notebooks contain an unusually rough sketch of a fleeting idea for a parachute, a pyramidlike, four-cornered, wood-and-fabric structure with a man hanging from lines attached to each corner. An annotation explains its purpose: "If a man have a tent made of linen of which the apertures have all been stopped up, and if it be twelve braccia across and twelve in depth [in modern terms, an area of 70 square feet], he will be able to throw himself down from any great height without suffering any injury." In our day, we might find the design problematic; as one aeronautical engineer declared, it looks as though it should "fall apart or spin uncontrollably." It certainly did not resemble today's billowing hemispherical sail, with a small stabilizing hole at the top, and cords made of lightweight nylon. In 2000, however, a skydiver successfully descended more than 7,000 feet using a chute built to Leonardo's specifications before dropping the last few thousand feet to the ground with a conventional chute.

The Anatomist

The notebooks dating from around 1490 contain Leonardo's first significant anatomical drawings—the studies of the human skull. He borrowed techniques that he first learned while doing architectural studies a few years earlier. As Carmen Bambach, curator of Renaissance drawings at the Metropolitan Museum of Art, has commented:

> As an architect, he begins to think of buildings in terms of plans, sections, elevations, three-dimensional perspectival views of form. Now that sounds to us very normal, but that's basically Leonardo's legacy! He is doing this in architecture by 1490, precisely the moment when he undertakes his studies for the human skull, and guess what—he applies the same technique of the section, the dimensional view, and the elevation to create a consistent vocabulary for anatomical description.

This practice is typical of Leonardo. Methods used in one place are applied unexpectedly in another with startling results. The cutaway view, which he first used in architectural plans, such as the truncated drawing of a two-tiered stable, becomes an even more powerful tool when used to render the human anatomy. The multiple view, rendering the same object from different angles and elevations, and the layered symmetrical figure, each section displaying the object peeled away to a different depth, were both first used in drawings of buildings. Subsequently, they became indispensable tools for the study of anatomy for all later generations. These drawings are the beginning of a serious, thorough, and steady inquiry; over the following 20 years they combined Leonardo's scientific interests with his gifts as an artist to produce the most astounding achievements in the history of anatomy.

"Observe the motion of the surface of the water which resembles that of hair, and had two motions, of which one goes on with the flow of the surface, the other forms the lines of the eddies."

—LEONARDO

One other activity consumed much of Leonardo's attention after 1490 and similarly tested his capacity to bring together his disparate interests and abilities. In July 1489, the long-sought commission for the equestrian statue honoring Francesco Sforza was awarded to Leonardo. Satisfactorily completing this project required pushing the boundaries of science and engineering in support of art on a grand scale. It also offered him a recognized place at court, as well as personal apartments and new accommodations for his studio in the Corte Vecchia, the official residence of the young Duke Gian Galeazzo, nephew of Ludovico. It had taken years, but as the new decade began, the future seemed to hold more promise and opportunity than Leonardo had ever known.

Leonardo, "A Pair of Studies of a Human Skull,"
ca 1489. Royal Collection, Windsor Castle, U.K.

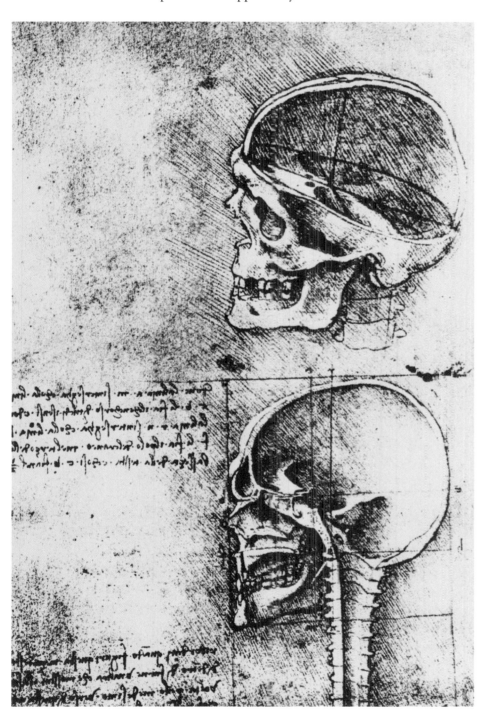

Leonardo's Sketchbook

The only book that Leonardo actually collaborated on and saw published in his lifetime was *De Divina Proportione,* published in Venice in 1509. The book, written by mathematician and friend Fra Luca Pacioli and illustrated by Leonardo, features 60 polyhedral figures, along with some architectural designs, and the design for a new font. The prominent letter "L" appearing in the frontispiece of this book is from that collection. The first two figures in this gallery are from the rare book.

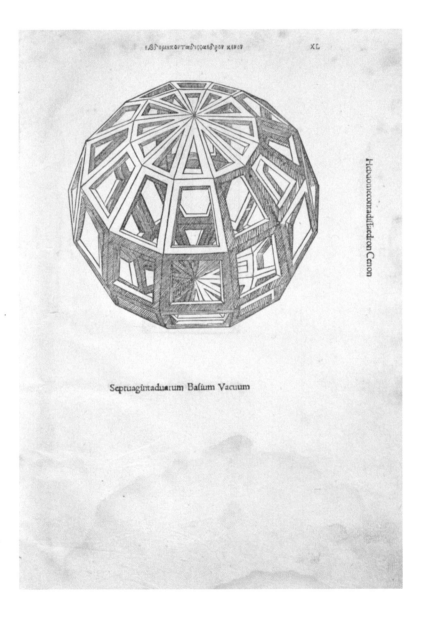

Illustration of a "72-sided sphere," with six rows of 12 faces each, from De Divina Proportione

Elevation of a portico, from De Divina Proportione

"Drawing of an Old Man and Geometric Recreation," 1489-1490. Royal Collection, Windsor Castle, U.K.

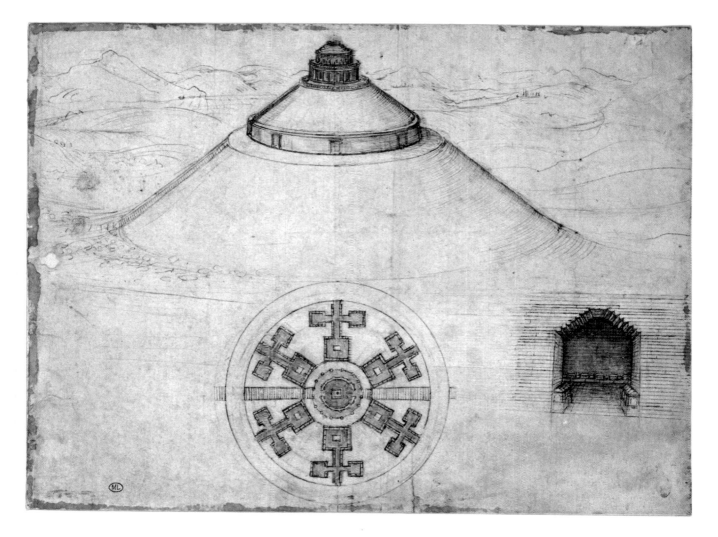

"Architectural Study for a Mausoleum," elevation and detail, ca 1504-08. Musée du Louvre, Paris

"A Study for a Figure for 'The Last Supper,'" ca 1495. Royal Collection, Windsor Castle, U.K.

"A Man Seen in Profile, with Geometric Figures Superimposed, in Order to Demonstrate Ideal Proportions," ca 1490. Galleria dell'Accademia, Venice

Chapter Four

"Human subtlety ... will never devise an invention more beautiful, more simple or more direct than does nature, because in her inventions, nothing is lacking, and nothing is superfluous." —LEONARDO

1491-99
THE MILAN YEARS

1490 Salai becomes apprentice to Leonardo.

1492 Christopher Columbus makes his first voyage across the Atlantic.

1493 Leonardo unveils clay model of Sforza horse.

1494 After yielding territory to the French, the Medici are exiled from Florence.

ca 1494-95 Ludovico Sforza commissions Leonardo to paint "The Last Supper."

1496 Leonardo begins collaboration with mathematician Fra Luca Pacioli.

1498 Fra Luca Pacioli dedicates his treatise *De Divina Proportione* to Ludovico Sforza (it is published in Venice in 1509). Girolamo Savonarola is executed for sedition and "religious errors." Niccolò Machiavelli, former patron of Leonardo, becomes Chancellor of Florentine Republic.

1498 Leonardo completes "The Last Supper."

1499 Ludovico Sforza is forced out of Milan. French troops storm the city. Leonardo leaves Milan.

With the commission for the Sforza horse and the new accommodations provided him at court, Leonardo achieved the security and comfort he required to do his best work. Indeed, 1490 and the years immediately following were among the most prolific and creative of his entire career. He was now a recognized member of the Milanese court, designated as an *ingeniarius ducalis*—a master creator for the duke—of which there were officially 13 at the time, including one of Leonardo's friends, the architect Bramante. Soon he had more commissions and assignments than he could handle.

At Court

Leonardo became chief set and costume designer for court theatricals—plays, masques, pageants, operettas—and apparently, judging from the testimonials that praise his spectacular sets and special effects, he was very good at it. When Ludovico "Il Moro" Sforza staged a gala in the Castello Sforzesco called "The Feast of Paradise" in January 1490, Leonardo was responsible for the conclusion to the evening, called "The Masque of the Planets"; in this mini-production, elaborately costumed actors, representing the five known planets of that time, moved through a model of the zodiac in their assigned courses by means of Leonardo's machinery, the entire production illuminated by torchlight. Some of Leonardo's most ingenious machines, such as "the car" and the humanoid, robotlike automaton in knight's armor, were probably invented for such court occasions. In 1491 he even organized, staged, and costumed an elaborate joust involving Ludovico's son-in-law, Galeazzo Sanseverino, in which men on horseback were dressed in wild outfits and Galeazzo carried a shield featuring a barbaric, bearded man.

In the late 1480s and early 1490s—the period of Leonardo's famous grotesques—he made bizarre, exotic, fantastic, threatening, and often ugly drawings of old men, women, and animals. Some of the drawings were probably studies

Opposite: Leonardo, "Portrait of a Lady" (aka "La Belle Ferronière"), ca 1495. Oil on wood. Musée du Louvre, Paris. Previous pages: Traffic streaks through central Milan as evening falls.

the ROBOTICS

Above: A reconstruction of Leonardo's robot knight of 1495. In 1996, Mark Rosheim first offered compelling evidence that Leonardo had designed robotic automatons.

Opposite: Leonardo, "Plans for a Programmable Moving Platform Powered by a System of Springs and a Differential Transmission," 1478-1480. Codex Atlanticus, Biblioteca Ambrosiana, Milan

> "His genius for invention was astounding, and he was the arbiter of all questions relating to beauty and elegance, especially in pageantry."
>
> —PAOLO GIOVIO

Leonardo combined his interests in mathematics, anatomy, sculpture, drama, physics, and mechanics to create some of his most interesting inventions, the automatons, or what we today might call robots, including the "automobile"—a self-propelled and programmable moving platform—and the now famous knight in the suit of armor, which could bend its legs, sit up, open its visor, turn its head, open its mouth, and make noise, all in a convincing lifelike manner.

Leonardo's earliest encounters with such devices and their underlying mechanics surely took place in Verrocchio's workshop, where, Vasari tells us, Verrocchio designed an automaton clock that astonished everyone. We do not know exactly what contribution Leonardo may have made to the design of such products while he was associated with Verrocchio, but many of Leonardo's drawings related to his automatons resemble nothing so much as the working parts of clocks and watches. Not long after he had opened his own workshop in the late 1470s, Leonardo designed his so-called automobile, or car (see opposite page). The drawings on this folio from the Codex Atlanticus puzzled scholars for centuries. In 1929 Guido Semenza first proposed that the drawings represented a self-propelled vehicle. And in 1975, Carlo Pedretti convincingly argued that the vehicle was also a programmable automaton whose movements—time, rate, and direction—could be pre-encoded into the device. These ideas are now generally accepted.

In April 2004 a working model of the car went on display in Florence at the Museum of the History of Science. Mark Rosheim, a NASA scientist and robotics expert, building on the insights of Carlo Pedretti, and using other surviving drawings in the Codex Atlanticus, demonstrated the high probability that the car was an accurate representation of Leonardo's intentions—that it was in fact a working machine, what Rosheim claims was "the first record of a programmable analog computer in the history of civilization," and not just an idea, such as his sketch of a human-powered helicopter, for future inventors to refine and improve upon. The car was almost certainly used as a platform in pageants, masques, banquets, and other productions at court to move players or parts of the set around the stage at predetermined points in the drama, or to produce other "special effects" that no doubt dazzled the audiences of that time.

Rosheim also reconstructed a model of the robot knight that had premiered in Milan in 1495. Leonardo's surviving drawings depicting the mechanisms of the knight are few. Many, if not most, of his sketches for this device are lost, among the thousands of pages bequeathed by Leonardo to Francesco Melzi, scattered across Europe by Melzi's heirs, and never seen again. But some of these now lost pages were seen by a select few before disappearing into oblivion. Rosheim notes that there is evidence that a Leonardo manuscript on biomechanics was in circulation as late as 1600 and most likely influenced Giovanni Alfonso Borelli's *De Motu Animalium (On the Motion of Animals)* published around 1675. Borelli held the chair in mathematics at the University of Pisa and like Leonardo was interested in the mechanics of flight as well as the mathematical description of human and animal motion. Using Borelli's work, especially plates from the *De Motu Animalium*, to help him interpret drawings from the Madrid Codex I, Rosheim was able to reconstruct a probable working model of Leonardo's knight.

Reconstructing the car and the knight from sketches left by the maestro was a task of interpretation, for Leonardo left few accompanying words next to these drawings. They are both examples where, as Rosheim says, "drawing for Leonardo is superior to words in communicating an idea."

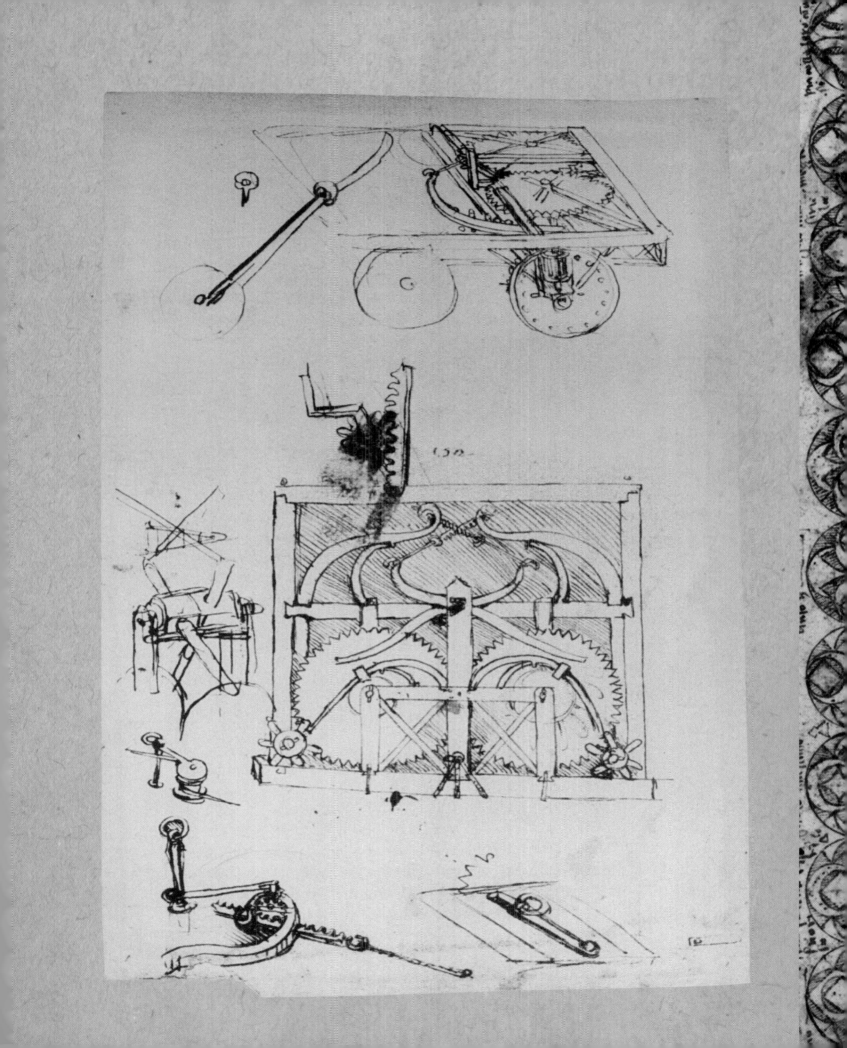

for these court productions; some were no doubt caricatures of real people he encountered in the streets; and some were inventions of his imagination, dwelling on deformity, ugliness, and the ridiculous. They are wild exaggerations, sometimes humorous and cartoon-like, the very opposite of the perfectly ordered and proportioned "Vitruvian Man" of the same period, but also a testament to his growing powers of observation and psychological insight.

By the early 1490s Leonardo may have been at work on another portrait, another work that today has disputed provenance. Like the subject of "Portrait of a Musician"—in which Leonardo most likely painted the subject's face and hair but

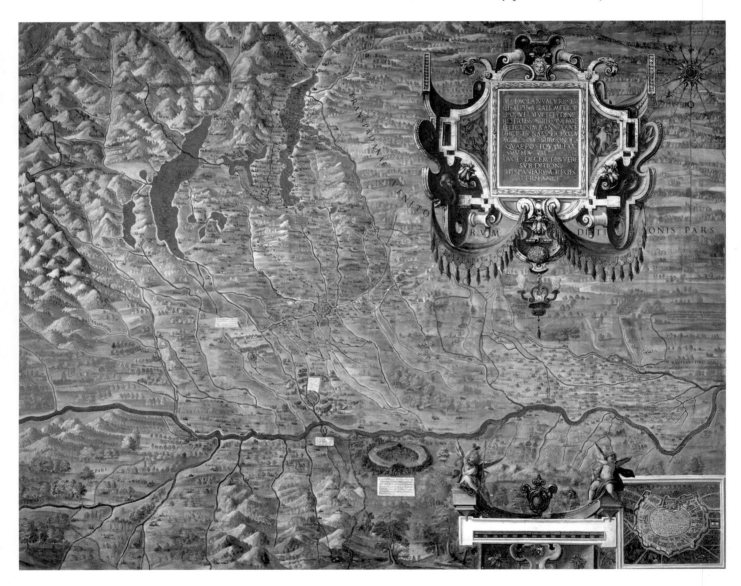

not his tunic or the portrait's background—questions also surround "Portrait of a Lady," also known as "La Belle Ferronière." Who is the subject of the painting, and who is the painter?

Among possible candidates for the lady depicted in the portrait is a grown-up Cecilia Gallerani, subject of "Lady with an Ermine." But this is unlikely, since there is no resemblance between the two portraits. A second candidate is the Queen of Spain, Isabella of Castile and Aragon, who sponsored Christopher Columbus's explorations. But Leonardo never visited Spain, and Queen Isabella is not known to have visited Leonardo. A third candidate is Madame Ferron, the wife of a prominent Parisian lawyer and the mistress of the French King, François I (1494-1547), Leonardo's last and most loyal patron. But the painting was completed by the mid-1490s, about the time François was born and 20 years before Leonardo went to France under his protection.

The prevailing theory among Leonardo scholars regarding the identity of the subject in "Portrait of a Lady" focuses on Lucrezia Crivelli, successor mistress to Cecilia Gallerani in Ludovico Sforza's affections. The painting's more familiar title, "La Belle Ferronière," derives from the thin hair band, or *ferronière,* worn across the subject's forehead, fashionable among Milanese ladies at the time. Painted in oil on walnut board, the portrait reflects a style introduced by Flemish artists—in which only the head and upper body of the sitter are visible with the hands of the subject hidden behind a banister or parapet seen at the lower edge of the painting.

Despite the prevailing theory as to the subject, nagging questions as to the painting's provenance still exist. Leonardo, since his apprentice days in Verrocchio's studio, had attached immense importance to the power of the hands. Would he have placed a wall in front of the subject, completely hiding her hands? The stiff demeanor of the subject and her heavy features also suggest that it may not be Leonardo's final product. Conversely, the subject's slightly averted glance and the fine braided rope around the neckline of the subject's dress are pure Leonardo. Further, this portrait shows the master artist's touch in depicting a Lombard lady who places great significance in displaying her most elegant clothes and jewelry.

A plausible guess as to the painting's provenance would be that Leonardo designed the posture and painted the face, the trim around the bodice, and the exquisite jewelry—and members of his studio collaborated on the rest. Ultimately, the portrait—with its conflicting clues—may or may not be from Leonardo's hand. Art historian Bernard Berenson expressed the dilemma eloquently: "[O]ne would regret to have to accept this as Leonardo's own work."

Salai

The early 1490s were also when Leonardo began what became the most important and enduring personal relationship of his life. Giacomo Caprotti, nicknamed "Salai—Little Devil," by Leonardo, became a member of the Milan household and studio in the summer of 1490. He was a ten-year-old from the town of Oreno, a few miles from Milan. Giacomo's father, Pietro, contracted with Leonardo to pay for the boy's upkeep in the studio, probably with the understanding that he would be trained as a painter. In fact, he became a barely competent painter under Leonardo's tutelage. Most of the works attributed to him are singularly undistinguished. During these first years with Leonardo, he served as an errand boy and artist's model. He was also a mischievous child, always in trouble, always up to something—thus the nickname bestowed on him. There is more about him in the notebooks than about any other character in Leonardo's life. He recorded the

Above: Leonardo, Detail from a drawing of an old man and a youth, most likely Salai, ca 1495. Galleria degli Uffizi, Florence

Opposite: Egnazio Danti, "The Duchy of Milan," including a map of the city of Milan, ca 1580. Fresco. Galleria delle Carte Geografiche, Vatican Museums

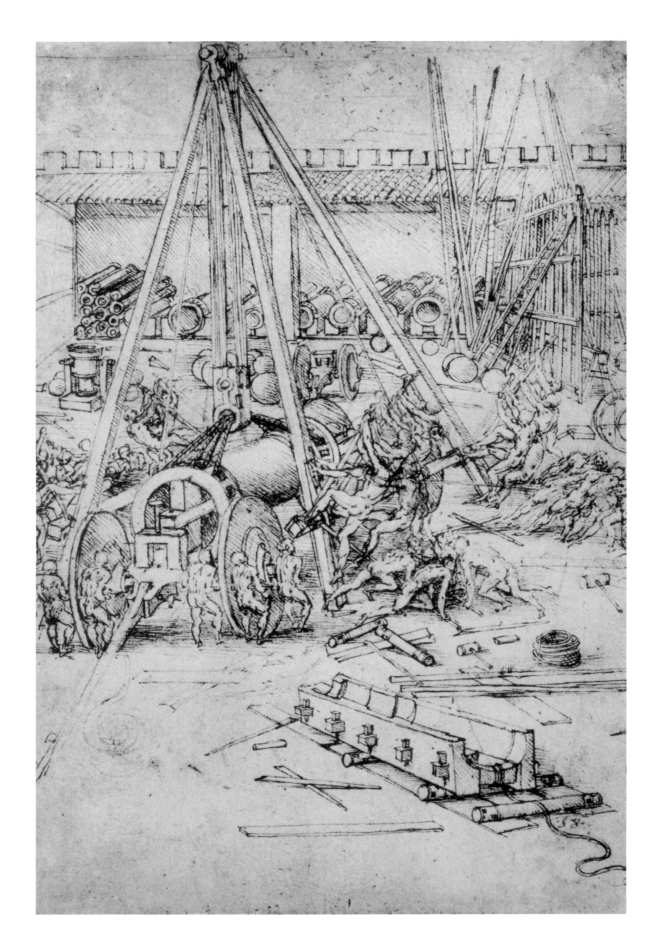

boy's miscreant behavior—stealing equipment, personal effects, and money left about the studio, breaking valuable dishes and spilling bottles of wine at the homes of friends and acquaintances. Leonardo kept careful records of money spent on the boy's clothes, food, and entertainment. These expenses continued into Salai's young adulthood. But the tone is rarely one of genuine complaint. Through all Leonardo's remarks about the boy—and the young man he became—runs a current of tolerance and fondness.

Salai remained part of Leonardo's life and entourage for the next 28 years and was treated generously in Leonardo's will; he received the house and garden just outside the walls of Milan given to Leonardo by Ludovico Sforza. Vasari tells us that Salai was "extraordinarily graceful and attractive" and with the ringlets that so delighted Leonardo. And Lomazzo, one of the four early Leonardo biographers, states more unequivocally that Salai became a regular sexual partner of Leonardo. Until recently, most Leonardo biographers refused to acknowledge the probability that Leonardo's relationship with Salai became a sexual one. Sexual or not, it was the most intimate relationship of Leonardo's life. The notebooks speak of quarrels and reconciliations, doubts and reassurances, confidences and betrayals—what we might expect of such a close association. Salai remained at Leonardo's side almost to the very end.

It is possible, indeed probable, that not long after the arrival of Salai, Leonardo's mother, Caterina, came to reside with him in Milan. He records in a notebook that "Caterina came on the 16th day of July 1493." Who was this Caterina if not his mother? Some biographers have speculated that it might have been a servant girl with the same name as his mother. This is unlikely. Though Leonardo never refers to her as "mother," she is mentioned by name, just as he mentions Salai, in his accounts of money spent in the upkeep of his household and its members. And when Caterina dies in 1495, when she would have been in her late 60s, he provides for the expenses of her funeral, which he carefully records. He would not have done this for a servant girl. But Leonardo leaves no clues about how Caterina's presence affected him psychologically. There is only the statement that she came to Milan, and the evidence that she lived there in her last few years at Leonardo's expense.

Futuristic Weaponry

The pacifist Leonardo left sketches for an array of war engines, some defensive and some offensive—not only to resist invaders but also to attack enemies. By the 15th century, the art of weaponry had been advancing for many years; some of Leonardo's weapons are derived from earlier times, but many of his drawings are entirely original, prefiguring technology of the distant future. Whether derived or original, all display innovations to assure efficiency, portability, and potency. In his famous letter of introduction to Ludovico, Leonardo wrote: "I can make armored cars, safe and unassailable, which will enter the serried ranks of the enemy with their artillery, and there is no company of men at arms so great that they will break it." With the armored car, or tank, breaching enemy lines, he continued, "the foot soldiers will be able to follow quite unharmed and without any opposition."

"I went to supper with Giacomo Andrea, and [Salai] ate for two, and did mischief for four, in so far as he broke three table-flasks, and knocked over the wine."

—LEONARDO

"I also have models of mortars that are very practical and easy to transport, with which I can project stones so that they seem to be raining down; and their smoke will plunge the enemy into terror, to his great hurt and confusion."

—LEONARDO

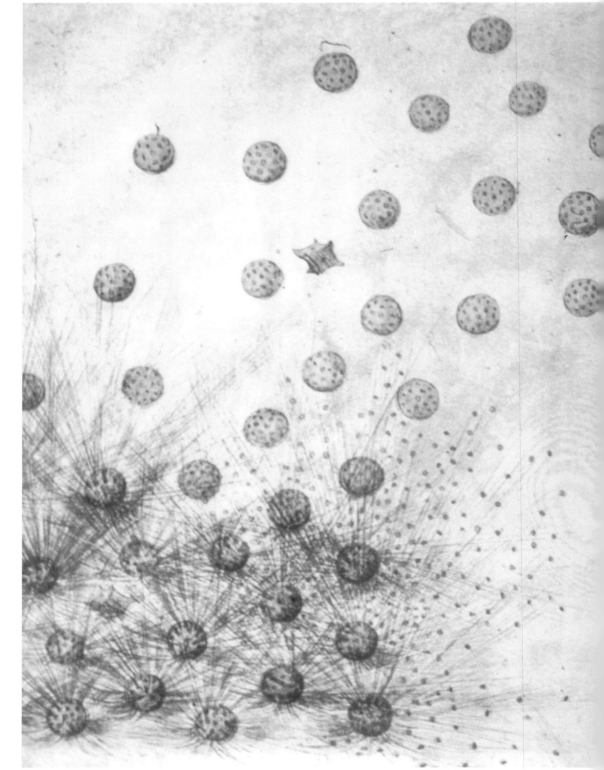

Leonardo, "Design for Shrapnel Mortar," ca 1495-99. Codex Atlanticus, Biblioteca Ambrosiana, Milan

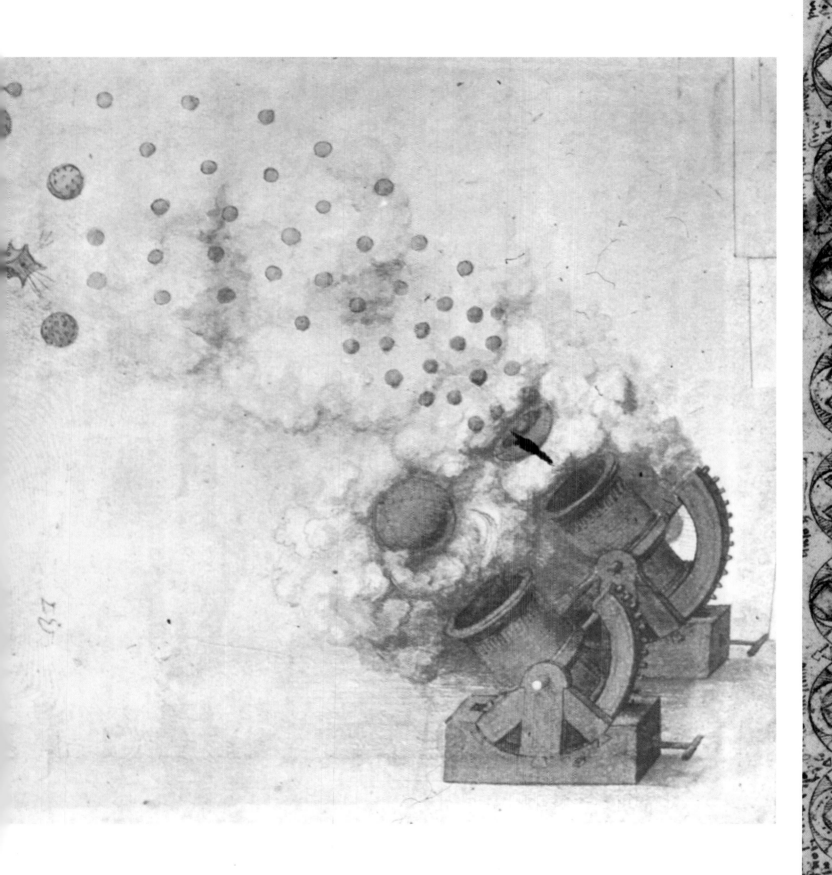

On the Italian Peninsula, the collective memory included Hannibal's elephants, which 1,700 years earlier had trampled their way through Roman lines. Leonardo was offering his own elephant, a thick, disc-shaped vehicle, featuring tapered surfaces clad in steel plates to deflect enemy fire. The inspiration, as with so many of his innovations, came from nature. The land turtle, or tortoise, nature's own armored creature, possesses a heavy shell, or carapace, and served as the model for Leonardo's tank. In Leonardo's design, a team of four or eight men applied torque in concert on a pair of cranks to power the tank. In creating the design, he provided both the external and the cutaway image to show the details of the

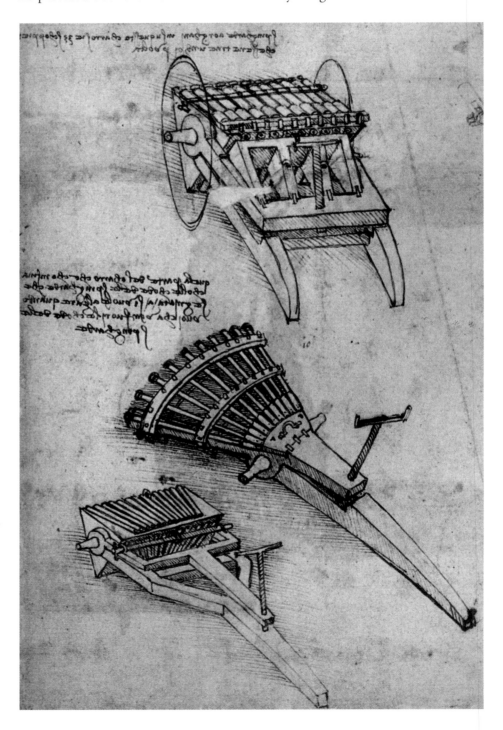

vehicle's power train. A scrutiny of the latter reveals a subtle design flaw: When the cranks are turned, the front and rear wheels work against each other. This could be just a simple mistake in what is obviously a very rapidly executed drawing. Some scholars have suggested that Leonardo may have introduced the flawed design into the drawing for other reasons, either to keep its true design a military secret secure from the enemy, or motivated by his pacifism, to keep such a deadly weapon off the battlefield.

In 1903 futurist H. G. Wells published a science fiction story, "The Land Iron Clads," in *Strand* magazine, inventing again the concept of an armored military vehicle. The first practical tank was manufactured and deployed by the British in September 1916 in World War I in the Battle of the Somme. Leonardo's tank had to await the development of a number of essential elements: the internal combustion engine and caterpillar tracks.

Though never leading to the actual production of equipment, the military drawings were a valuable exercise for Leonardo. Even the unorthodox catapults and the repeating crossbow, though looking to the past, were for Leonardo the start of his encounter with complex devices and their relation to the forces of nature. In rendering these engines, Leonardo refined his mechanical drawing until it was ready for the careful depiction of the complicated geared machinery he invented in the 1490s, the kind of machines that would power the industrial revolution 350 years later.

Included among Leonardo's designs was a daunting arsenal: a front-loading firearm that had an adjustable gear with a peg-blocking system, intended to be used by soldiers; an adjustable light-caliber gun with eight muzzles mounted on a two-wheeled carriage; a 33-barrel machine gun, mounted on a revolving framework, which could fire 11 shots simultaneously, and as one set of 11 was fired, a second set would be loaded by trained artillery crew. As for gunpowder, Leonardo developed his own recipes, and he designed shells that exploded upon impact and others that exploded in mid-air into smaller fragments—shrapnel.

One device that was produced in Leonardo's own lifetime was the wheel lock. As a precursor to the flintlock, the wheel lock represents a significant development in the history of weapons, making possible portable guns—pistols, muskets, the blunderbuss, and ultimately the modern rifle. Leonardo's designs for a door lock and a wheel lock both appear in the pages of the Codex Madrid I; his designs for a variety of leaf and coil springs appear in the Codex Atlanticus. Springs and small connecting chains are used in cocking and releasing component parts. When a trigger is pulled, a cocked mainspring sets a steel flywheel spinning. A piece of iron pyrite—a flint—held in place by a small vise is brought into contact with the rim of the spinning wheel, resulting in the release of a stream of sparks. These ignite the gunpowder interposed in the barrel between the wheel lock and a bullet—and the design also calls for a stock heavy enough to reduce the recoil to the gun bearer's shoulder, allowing the weapon to be held safely and comfortably and aimed with ease. As conveyed in the common expression "lock, stock, and barrel," Leonardo invented all the necessary components of the modern rifle. In his notebooks, designs for the rifle appear on the same sheet as his scythed chariot, a frightening contraption meant to mow down attackers during a siege.

"I will make covered vehicles, safe and unassailable, which will penetrate enemy ranks with their artillery and destroy the most powerful troops; the infantry may follow them without meeting obstacles or suffering damage."
—LEONARDO

Following pages: Leonardo made frequent visits to the Dolomites, where he would have glimpsed Monte Cristallo, located in today's Fanes-Sennes-Braies Nature Park.

The Consummate Scientist

The Milan years were the first years of Leonardo the serious scientist—as his fascination with nature took a deliberate and organized turn toward careful research and observation. In the spring of 1490 he began a small notebook devoted almost exclusively to a single line of scientific inquiry—optics, or the behavior of light. Careful and neat throughout, the drawings are painstaking, meticulous, and they incorporate his knowledge of geometry. These are tantamount to modern lab books.

These years are also the time of his trips to the mountains. Leonardo did not travel widely. He never left Italy until he departed for France near the end of his life, never to return. But he did take to the road when he could, especially if presented with an opportunity for unusual observations. In the early 1490s, he twice headed north into the Italian Alps. He took a notebook and jotted observations and made sketches along the way, noting additional speculations and drawing more elaborate sketches from memory in the quiet of his study after his return. Plants, animals, and, as always, the birds were carefully described and recorded. And he takes note of the people of the region and how they have adapted to their austere environment. These treks resulted in some remarkable drawings of thunderstorms that accurately depict the direction and force of the violent downdrafts known as wind shear. He also examined fossils, rock strata, and other land formations, and made some of the first modern, scientific speculations about geology and natural history. His drawings of cliffs show clearly the strata from different geological epochs and large rocks thrust up from the Earth's crust. And most dramatically, he speculated about a much older origin for the Earth than the conventional view.

Science and technology, even in our own age, are mistakenly regarded as being synonymous. Certainly in modern times progress in each has been the result of symbiotic collaboration with the other. Historically, however, the two fields have different origins and have served different purposes. Technology involves the creation of tools to make physical tasks easier and to improve the quality of life, and dates back to the earliest humans living on the savannas of Africa. From the time they learned to sharpen stones to use as implements for cutting and impaling, or learned how to start fires for cooking food or smelting ore, these early humans were engaged in technology.

Science, on the other hand, seeks to identify the laws of nature, and with these to explain natural phenomena. It is more abstract and its origins date to the pre-Socratic philosophers of ancient Greece. Thus science is a mere infant. Thales of Miletus (ca seventh century B.C.), regarded as the first scientist, is credited with the timeless pronouncement, "One does not have to attribute natural phenomena to the whims and vagaries of the Gods, but rather to look for causes through natural law."

Classical Mechanics

Classical mechanics, a subfield of physics, deals with force, mass (weight), acceleration, and velocity in describing the motion of finite objects. Galileo in the late 16th and early 17th centuries discovered the law of free fall (that is, free of drag), in which bodies of different masses (weights) fall with constant acceleration. This

"We may say that the Earth has a vital force of growth, and that its flesh is the soil; its bones are the successive strata of the rocks which form the mountains; its cartilage is the porous rock, its blood the veins of the waters. The lake of blood that lies around the heart is the ocean. Its breathing is the increase and decrease of the blood in the pulses, just as in the Earth it is the ebb and flow of the sea."

—LEONARDO

Opposite: Leonardo, "Storm in the Alps," ca 1503-05. Royal Collection, Windsor Castle, U.K.

-137-

Law of Free Fall

Galileo, in the 17th century, found that a streamlined object (shaped to minimize drag), when dropped from a tall tower, displaces 1, 3, 5, 7,... units of distance in sequential one-second intervals. This led him to conclude that acceleration of a falling body is constant. Leonardo, with whatever crude timer he used, found a falling body to displace 1, 2, 3, 4,... units of distance in the same one-second intervals. He also arrived at the same conclusion of constant acceleration. Although Galileo spent several decades studying and teaching in Pisa, there is no evidence that he used the Leaning Tower for his experiment, and certainly none that Leonardo performed his experiment there, but the tower offers a convenient shape to illustrate the law of falling bodies. In reality, Galileo arrived at his law of free fall by allowing smooth balls to roll down inclined boards, thereby reducing their acceleration relative to falling bodies. He then extended his idea to the latter.

The famous bell tower of the Duomo of Pisa, known more commonly as the Leaning Tower. Built during a 200-year period commencing circa 1160, the tower's angle of inclination of 5.5 degrees is especially conducive for the illustration of the law of falling bodies.

signaled a clear break with the prevailing Aristotelian physics of his time, which had claimed that heavier objects accelerated faster than lighter objects. Isaac Newton in his incomparable 1687 book, *Philosophia Naturalis Principia Matematica (Mathematical Principles of Natural Philosophy)*, first introduced the three laws that bear his name, as well as the universal law of gravitation, and showed why acceleration was indeed constant for bodies of different masses. Two hundred and twenty-eight years after the *Principia*, Albert Einstein published his own masterpiece, the theory of general relativity, and introduced the currently prevailing theory of gravitation (with four-dimensional warped space-time).

The quest to understand gravitation involves a short list of the greatest scientists in history. It should not surprise us that Leonardo also grappled with gravitation, and wrote of constant acceleration, more than a century before Galileo. He also anticipated Galileo in describing projectile motion. It is well known in classical mechanics that the trajectories of projectiles—whether arrows, cannonballs or

basketballs—are parabolas. Aristotelian physics, still being taught in artillery schools well into the 17th century, incorrectly described projectile motion: The cannonball would rise in a straight line at some oblique angle, gradually curve downward, and then plummet straight to the earth. Galileo in 1609-1610 established the trajectories of objects sliding off horizontal surfaces as being parabolic. Leonardo, with his astonishing observational capability, had sketched out parabolic trajectories as early as the 1490s. So astute are his observations on the trajectories of the cannonball that he is able to comment that the ratio of the apogee (maximum height)

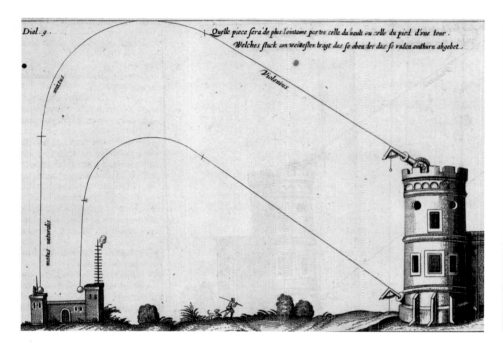

to the range (horizontal displacement) is preserved in successive bounces of a cannonball striking the ground.

Leonardo was both a pure scientist and a practical and prescient technologist. Much like scientists of the future such as Galileo, Newton, and Einstein, he designed and constructed equipment specifically to assist his scientific inquiries into physics. There are a number of areas in pure physics in which he prefigured Galileo, and even anticipated one very significant invention by Newton.

Leonardo worked in optics from the vantage point of both the artist and the scientist. With keen observation, he realized as early as the 1470s that, as light traveled through the air, with increasing distance the quality of light would be degraded. Even in his earliest paintings and drawings, he revealed this understanding. After studying light refracted through glass lenses, which he presents with ray diagrams, he alludes in several codices—Atlanticus, Arundel, and Madrid II—to "making glasses to see the moon enlarged." He appears to be prefiguring the refractor, a telescope eventually invented by the Dutch optician Hans Lippershey in 1608, and improved and put to astronomical use by Galileo just a few months later. Then, in 1513, after studying light rays reflected from concave surfaces and seeing that they converge at a single focal point, Leonardo notes: "In order to observe the nature of the planets, open the roof and bring the image of a single planet onto the base of a concave mirror. The image of the

Above left: Reproduction of a drawing regarding trajectories that first appeared in Diego Ufano's Artillerie, *published in 1621.*

Above: Leonardo, "Depiction of Parabolic Trajectories in Projectiles Launched at Different Angles," 1487-1490. Codex Madrid I, Biblioteca Nacional de España, Madrid

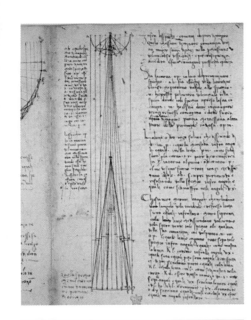

planet reflected by the base will show the surface of the planet much magnified." With this device he was foreshadowing Isaac Newton's 1668 invention of the reflecting telescope. The multibillion-dollar Hubble Space Telescope, placed into orbit by NASA in 1986, one of the greatest scientific tools ever created, is a descendant of these early reflectors.

Among all of the areas of natural science, Leonardo made the most dramatic strides in the life sciences. The quality of his anatomical studies far exceeded those published by Vesalius in 1543 that would revolutionize medicine. Of course, one of the tragedies of Leonardo is precisely that his discoveries were not published and never became seminal to the works of future scientists. His studies in physics were carried out mostly in the 1490s, but his best work in anatomy would come two decades later. An earth-shattering pronouncement on science comes in a slightly ambiguous statement where he appears to offer his thoughts about the evolution of life itself, prefiguring Darwin by 350 years: "Nature, being inconstant and taking pleasure in creating and continually producing new forms, because she knows that her terrestrial materials are thereby augmented, is more ready and more swift in her creating than is time in his destruction."

The Sforza Horse

In the years immediately after 1489, work on the magnificent statue of Ludovico's father atop a horse was Leonardo's artistic priority. It began with drawings trying different conceptions of the statue. His initial vision was bold—and impossible. Leonardo wanted to have the horse rearing on its hind legs, but he had to concede that, for a statue of this size, the hind legs would not be strong enough to balance and support the rest of the horse.

The full-scale clay model of the horse that Leonardo eventually constructed did not survive, and it was never cast in bronze as planned. We do not know exactly what might have been his final vision for the statue. But the progression of surviving drawings in the Codex Atlanticus indicate that he settled at last on the idea of a trotting horse. Once he had decided on the pose and appearance of the horse, his next step was to construct the full-scale clay model needed to create a mold. The lost wax process that Leonardo planned to use to cast the horse called first for a mold of the clay model, then a layer of wax on the inside surface of the first mold, then another mold on the inside surface of the wax. After the molds hardened the wax would then be melted away and replaced by molten bronze. Nothing of this size had ever been cast in bronze before using the lost wax—or any other method. Leonardo, as was so often the case, found himself in uncharted territory. It was a challenge just to accumulate the necessary bronze—more than 70 tons.

Leonardo decided to cast the horse in one pour of molten bronze instead of pouring sections of the horse separately. This required the finished mold to be put, lying on its side, into a pit, instead of upside down into a much deeper pit. This way the bottom of the mold would not be in contact with the very shallow water table beneath Milan, causing uneven cooling of the molten bronze. This plan necessitated the design of machinery to maneuver the mold and the finished horse—cranes, hooks, pulley systems, and other lifting devices. The drawings for these devices are the beginning of Leonardo's intense engagement with the working of complex combinations of gears, chain drives, axles, universal joints, screw threads, and roller bearings. Codex Madrid I, begun in 1493, around the time Leonardo was making plans for the casting of the horse, contains some of Leonardo's first examples of fine mechanical drawing, an art he raised to a new level. In November 1493, Leonardo publicly exhibited his finished clay model on the occasion of the marriage of Bianca Maria Sforza, Ludovico's niece, to Holy Roman Emperor Maximilian. The sculpture, about 24 feet high, was placed in

"Those who saw the great clay model that Leonardo made considered that they had never seen a finer or more magnificent piece of work."

—VASARI

front of the Castello Sforzesco and praise came from all directions. Bramante commented cleverly: "*Vittoria vince e Vinci tu vittore*—Victory to the victor, and you Vinci have the victory."

The moment of Leonardo's greatest triumph seemed at hand. But politics were about to interfere in the fate of the sculpture and the life of Leonardo. After the death of Lorenzo de' Medici of Florence in 1492, the delicate balance of power that had preserved the peace of the Italian Peninsula for the previous 50 years began to unravel. Longtime alliances among the kingdoms and city-states were broken and new combinations were formed, but no stable arrangement could be found. Eventually King Charles VIII of France, on the premise of a weak claim to the Kingdom of Naples, brought his armies into Italy and to the gates of Milan. Ludovico, now officially Duke of Milan following the suspicious death of his nephew, Duke Gian Galeazzo, in 1494, needed the more

Above: Leonardo, "Study for the Sforza Monument," ca 1488. Royal Collection, Windsor Castle, U.K.

Opposite top: Leonardo's studies of the reflection of light rays from a concave surface, prefiguring the reflecting telescope. Opposite center: The first reflecting telescope constructed by Isaac Newton stands by his manuscript of the Principia. *Opposite bottom: Hubble Space Telescope, a reflecting telescope placed in orbit by NASA in 1986.*

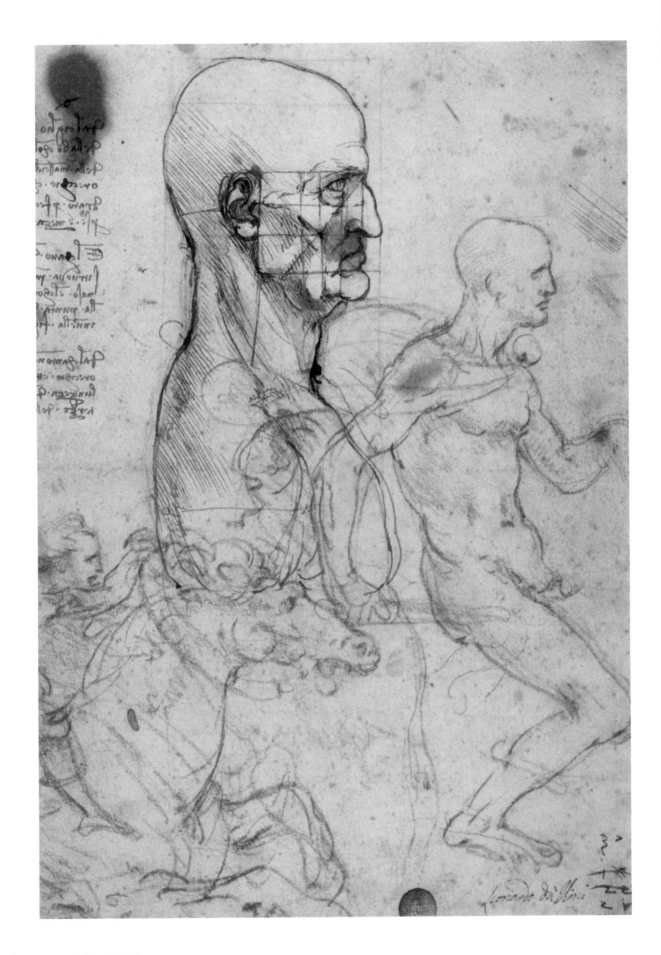

than 70 tons of bronze set aside for the horse to make cannon for the war effort. In 1495 the French were defeated and Ludovico survived as a weakened ruler. And Leonardo, resigned to the new reality, vowed to "speak of the horse no more."

A New Interest in Mathematics

In his Milan years, Leonardo became profoundly interested in geometry and algebra. He read Euclid's *Elements;* contemporary mathematician Fra Luca Pacioli's *Summa de Arithmetica, Geometria, Proportioni et Proporzionalita (Treatise on Arithmetic, Geometry, and Proportion);* Leon Battista Alberti's ten-volume *On the Art of Building;* and Piero della Francesca's *On Perspective for Painting.* In carrying out his own research in mathematics, he would frequently seek mechanical solutions to geometric and algebraic problems. No doubt it was in this period that he began to appreciate both the aesthetics of mathematics and the mathematics of aesthetics, and began to seek mathematical demonstration of natural laws. In such a quest he was revealing astonishing prescience—the Scientific Revolution of the 17th century was sprung by figures like Galileo, Kepler, and above all Newton who would express natural laws mathematically. For Leonardo, in this period, the urgency to produce art was subordinated to understanding mathematics, and painting became no more than a part-time endeavor.

Leonardo's patron in Milan, Il Moro, was known for his cruelty and ruthlessness, but as the regent he proved to be a surprisingly visionary leader. He championed the arts and sought to hire talented artists, writers, and musicians from outside, all with a view toward transforming Milan into a cultural rival of Medici Florence. While serving as regent, he hired Leonardo as court painter and engineer. His ascendance to the title of Duke came in 1494. Between 1494 and 1496 Luca Pacioli was invited to Milan as mathematician and tutor to the Sforza court. Leonardo, by then a resident of Milan for a dozen years, had become increasingly enamored of mathematics, and the invitation extended to Pacioli to come to Milan may well have been at his behest.

Fra Luca Pacioli was born in the Tuscan town of Sansepolcro in 1445 and died there in 1517. Despite these unremarkable biographical facts, Luca Pacioli was a kind of "itinerant mathematician" who seemed to know all the right people in both scholarly and ecclesiastical circles—and who significantly influenced Leonardo. A friar in the Franciscan order who had studied in Rome and Venice, Pacioli served as instructor at the Universities of Perugia, Zara (in Croatia), Naples, Rome, Pisa, Bologna, and Florence. During a five-year period from 1489 to 1494, Pacioli wrote his famous *Summa (Treatise).* As a compilation of all the mathematics formulated up to his time, it did not present much original work. Yet in a day when mathematics was strongly driven by the mercantile economy, he left an enduring legacy for modern accounting, specifically by introducing into Europe double-entry bookkeeping. Leonardo admired Pacioli immensely, and the two became good friends.

Leonardo's friendship with Pacioli allowed him to improve his knowledge of mathematics—but in giving himself to mathematics, he neglected his art. Indeed, Leonardo became so obsessed with the subject that a visitor to his studio in Florence described the artist as "out of touch" with painting. Guided by

THE REPLICAS

Isaac Newton's "mental inventions," his expression used to describe his abstract theories in physics and mathematics, may be even more appropriate to describe Leonardo's designs. The wheel lock did come to fruition, and proved to be seminal for future portable weapons, including rifles and handguns. A few others, such as the scissors, and the rolling wheelbarrow-like odometer—practical and simple to build—were most likely created in his time. But the vast majority of his other mental inventions, although accompanied by detailed specifications, were not built during his lifetime, and had to be reinvented much later.

During the late 1930s, Italian dictator Benito Mussolini commissioned engineer Roberto Guatelli to produce replicas from Leonardo's designs. Guatelli, staying true to the original specifications, produced operable models, which were subsequently shipped to Japan, to be exhibited in Tokyo. Shortly after the exhibition opened, however, World War II broke out, trapping the replicas, which were destroyed by the bombing of the city. After the war, Guatelli was hired by IBM, moved to the United States, and began again to produce a new collection of the machines. Later, IBM donated the collection to the Gallery Association of New York State (GANYS), with the understanding that the nonprofit institution would make the collection available for public viewing. Many collections of replicas now exist, built by a company that Guatelli founded, and by other groups operating independently.

MECHANICAL DRAWINGS

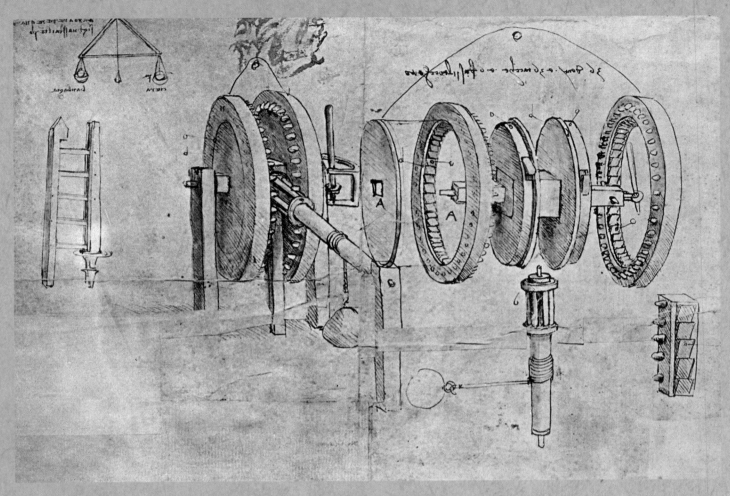

Above: Leonardo, "Exploded View of Gear Assembly and a Hygrometer (top left) for Measuring Humidity," ca 1505. Codex Atlanticus, Biblioteca Ambrosiana, Milan

Opposite: Leonardo, "Drawing of the Mold for the Head of the Horse for the Sforza Monument," ca 1491-93. Codex Madrid I, Biblioteca Nacional de España, Madrid

As a teenager, Leonardo was fascinated by the hoists and cranes of Filippo Brunelleschi, architect of the magnificent dome of the cathedral in Florence. Many of these devices were still on the site of the Duomo when Leonardo arrived in the city, and have been preserved for posterity in sketches done by the young studio apprentice. His first notebooks from the 1480s are filled with designs for machines and engines—from military equipment to hydraulic devices for lifting water. Most of these drawings, however, offer only enough detail—often with accompanying text—to demonstrate how the contraption would work.

But beginning in 1493, with the drawings that survive in the Codex Madrid I, Leonardo's technical studies took on a new level of precision and clarity. He was then at work on the Sforza horse, enmeshed in a number of technical problems relating to its casting. Most of these studies deal with the transmission of motion and power by the use of gears, springs, chain drives, and other such devices, usually without reference to any practical purpose, though the same sort of contrivances would eventually be found in the innards of the machines that would power the industrial revolution. The drawings demonstrate that Leonardo was engaged with natural forces and the principles by which they can be harnessed to do useful work. They are meticulously drawn, with perspective and shading, sometimes using the technique of the exploded diagram, which Leonardo may have invented. The often haphazard arrangement of previous notebooks has been replaced by an obvious concern for the appearance of the page. Perhaps he was preparing to publish something along these lines. Whatever his intentions, as he did in a number of other fields, he raised the science and art of mechanical drawing to a new level.

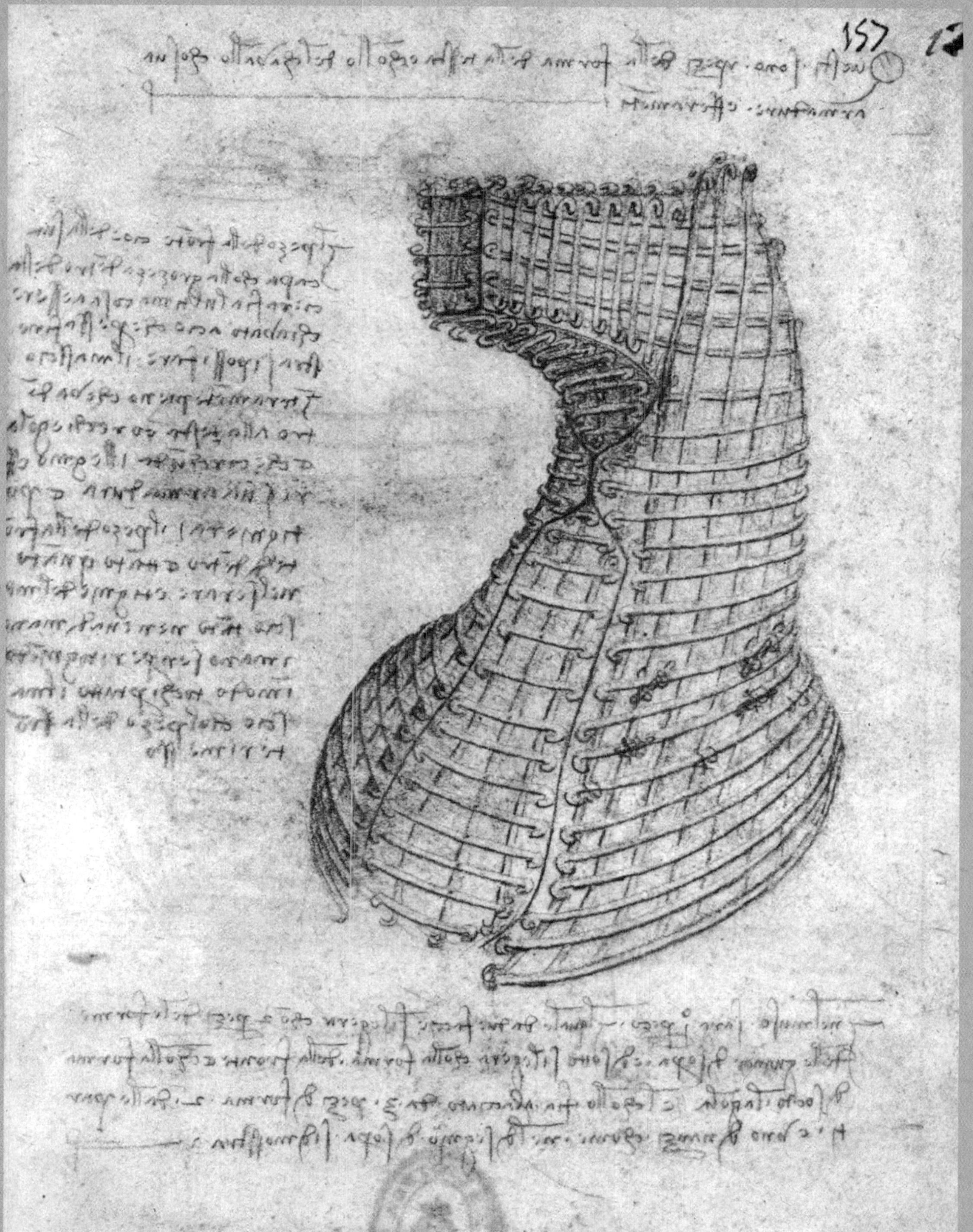
157

Pacioli, the 44-year-old Leonardo began to study the ancient authorities Euclid and Archimedes. The painter turned mathematician was preoccupied with a number of geometric problems, particularly that of "squaring the circle," or constructing a square equal in area to a given circle.

Leonardo's friendship with Pacioli led to their collaboration on *De Divina Proportione (On the Divine Proportion),* published in Venice in 1509. Pacioli's name appears as the author, and Leonardo's name is conspicuous by its omission. In the book Pacioli praises Leonardo for the beautiful designs he has made of geometric bodies—the pyramid, cube, octahedron, dodecahedron, and many

others. In his years in Milan, Leonardo undoubtedly mastered solid geometry, some of the shapes figuring prominently in his artistic compositions. Leonardo's designs, preserved in the original manuscript in the Ambrosiana Library in Milan, appear as woodcuts accompanying the printed text. Leonardo chose to render the complex three-dimensional bodies foreshortened and in pierced (transparent) form for greater clarity, the first time the polyhedral forms had been depicted in this manner. The abstract beauty of the polyhedrons obviously intrigued him, but their real significance lay elsewhere. In a sense Leonardo is both a neo-Platonist, believing in nature revealing itself through mathematics, and a neo-Aristotelian, believing that nature reveals its secrets through experimentation. In reality Leonardo was far more of an experimentalist than even Aristotle. The extremely rare book features 60 drawings by Leonardo—including a variety of regular and irregular polyhedrons, a drawing of a face on which is superimposed an equilateral triangle to illustrate proportions, and the design of a new font for the typesetter.

The Creation of "The Last Supper"

Ludovico's father, Francesco Sforza (1401-1466), as early as the mid-15th century, had commissioned the erection of a Dominican convent along with an adjoining church at the site of a small chapel, Santa Maria delle Grazie (St. Mary of the Graces), on the outskirts of Milan; the main architect was Guiniforte Solari. The convent was completed in 1469, while construction on the church continued for another two decades. In 1494 Ludovico, recently having assumed the title of Duke of Milan, decided to create within the complex the Sforza family tombs, emulating the Medici tombs in Florence. Donato Bramante was hired as the architect to finish the church; he finished the apse and dome before moving to Rome to become the architect of the flagship church of Christendom, St. Peter's Basilica.

The exact date of Leonardo's commission to decorate the north wall of the refectory, or dining room, of the friars is not known; the completion date of the mural, however, is documented as 1498. The theme, Christ's final supper with his disciples, was not uncommon for refectories of monasteries—and this one might have started out as the Duke's perfunctory present to the friars. It would prove to be a priceless gift to the worlds of art and religion. Leonardo's "Il Cenacolo" ("The Last Supper") is regarded as the first work of the Italian High Renaissance and is described by art historian Kenneth Clark as "the very keystone of Western Art."

Cosa Mentale

For Leonardo, with all his scientific, engineering, and painterly appetites and skills in full bloom, this work represented a *cosa mentale* (a thing of the mind)— an opportunity to synthesize the full spectrum of his passions. He was at once an artist who had perfected the use of sfumato, chiaroscuro, and perspective; an experimental chemist exploring the painter's media; a physicist who, after endless experiments with light and shade, had gained an intuitive understanding of optics; an anatomist who had been developing extraordinary expertise on the human body and its movements; a psychologist who had long studied facial and hand gestures. Now in the last few years the friendship developed with Luca Pacioli had given him insight into mathematics, which was increasingly finding its way into various facets of his intellectual endeavors, including his art.

Leonardo decided to customize the painting so that it would appear to be an extension of the room, as if the disciples were sitting in the refectory. The natural

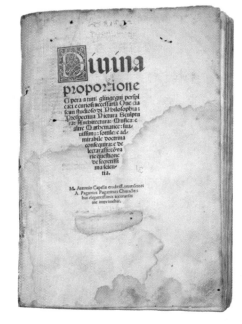

Above: Title page of Luca Pacioli's De Divina Proportione, *illustrated by Leonardo*

Opposite: Jacopo de' Barbari, "Portrait of Luca Pacioli," ca 1495. Museo di Capodimonte, Naples. A glass model of the rhombicuboctahedron is suspended on the left, and a dodecahedron sits on a book on the right. Controversy surrounds the young man in the portrait: Is he a self-portrait of the painter, a depiction of the young Albrecht Dürer, or the son of Ludovico Sforza?

"Each of the men at the table reveals himself in an instantaneous psychological portrait that seems to betray his thoughts and even those thoughts that will come later."

—SHERWIN NULAND

LEONARDO DA VINCI, 2000

During World War II, Santa Maria delle Grazie, in Milan, received a direct hit by an Allied bomb that demolished a wall at right angles to that bearing Leonardo's "Last Supper." The fresco, heavily barricaded with sandbags, sustained minimal damage.

light enters the room through windows located high on the western side of the wall designated for the painting. With this in mind, Leonardo planned to illuminate the right side of the scene far more than the left, and also to illuminate the left sides of the subjects, while their right sides would be seen in gradations of shadow. "The Last Supper" is a large mural, 30 by 14 feet, elevated high above floor level. It incorporates impeccable one-point perspective, the horizon line passing through Christ's head, and all of the lines of perspective converging at a vanishing point on Christ's forehead. Leonardo painted three windows on the wall behind the disciples, the center window serving as a backlight for Christ, creating a natural halo.

Earlier Renaissance artists, in depicting Christ's last meal with his disciples, had almost always separated Judas from the others, placing him alone in the foreground. His face was never presented, lest the viewer gaze inadvertently into the eye of evil. Accordingly, he was seen from the back, facing the other disciples with Christ seated across from him. In the earliest of his studies for the mural, Leonardo apparently considered bowing to tradition and separating Judas from the others. Also, St. John, a favorite, was usually seen with his head resting on Christ's shoulder or even on his lap. And in Leonardo's preliminary studies one could see him trying out these traditional approaches, but clearly trying to capture the mood at that dinner table, experimenting with the facial expressions and hand gestures of each of the characters. With the years of relentless sketching of strangers on the street and in the marketplace, he had compiled a storehouse of diverse faces and characters. Leonardo achieves heightened drama by casting the characters in four groups—two trios on the left of Christ, two trios on the right. St. John is leaning away from Christ, completing the trio just to the left. The model is a somewhat effeminate character, and one that Leonardo will use later in his portrait of St. John the Baptist.

In Leonardo's "Last Supper," starting from the left, the first trio comprises Bartholomew, James the Younger, and Andrew; the second trio, Judas Iscariot, Peter, and John. Jesus is symmetrically isolated from the other 12 men. The third trio consists of "Doubting" Thomas, James the Elder, and Philip. The fourth trio presents Matthew, Judas Thaddeus, and Simon.

By 1495, when Leonardo finally began to apply his paint on the wall, he had rejected the notion of presenting Judas from the rear, settling on a more democratic lineup—the disciples were all to face the viewer. After all, he was depicting the instant after Christ has made his revelation that one of the disciples would betray him. Integrating Judas into one of the four trios, Leonardo placed Judas's face in shadow, showing just his profile.

In that electric moment, when Christ has just foretold of his betrayal, Judas recoils in terror, knocking over the saltcellar, but is not about to reveal that it is he who is the betrayer. The others react with mortified shock and disbelief. Leonardo employs the hands—as much as the faces—of the disciples to convey the powerful emotions of the moment. Following his own admonishment for young artists set out in the *Libro di Pittura (Treatise on Painting)*, Leonardo avoided the pitfalls often seen in multiple-subject works: "Do not repeat the same movements in the same figure, be it their limbs, hands, or fingers. Nor should the same pose be repeated in one narrative composition."

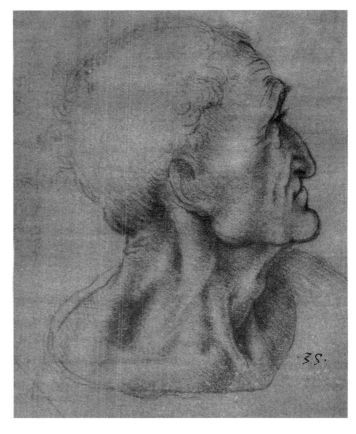

Leonardo's modus operandi in the creation of a piece of art is revealed by one contemporary observer's comment that Leonardo would stand in front of the mural and study it carefully for hours. Then he would sometimes pick up brush and paints, climb on the scaffolding to gain access to the mural, apply two or three strokes, and depart. Sometimes he would paint from sunrise to sunset, ignoring meals. Other times he would just ponder the work for hours, but leave without doing any painting. And after leaving, he would not be seen again for days. This was the pattern of the supreme perfectionist rather than the hopeless malingerer.

Leonardo, Vasari reports, characteristically running late in completing the work, was called to task by the prior, who complained to Leonardo's boss, Ludovico: "Why is it taking so long?" When Ludovico called Leonardo before him to convey the prior's impatience, Leonardo responded that if he could not find an appropriate face for Judas then he "could always use the face of the tactless and impatient prior!" Il Moro laughed uproariously, and the prior departed, fuming with indignation, never to badger the artist again.

Above: Leonardo, "Study for the Head of Judas in 'The Last Supper,'" ca 1495. Royal Collection, Windsor Castle, U.K.

Opposite: Leonardo, "Study for 'The Last Supper,'" 1495. Gallerie dell'Accademia, Venice

A MODEL FOR JUDAS

During the mid-1490s, when Leonardo was painting "The Last Supper" in Milan, a cultural revolution was sweeping Florence, threatening to destroy the city's cultural and artistic heritage. It had begun in 1492 when the Dominican friar Girolamo Savonarola came to Florence issuing condemnations and calling for the ouster of the ruling Medici family, whom he judged the major source of moral decay. Lorenzo's relations with the church had not been close, but they had been cordial and his promotion of humanism had never been a source of conflict. With Lorenzo's death in 1492 and the accession to power of his weaker brother, Piero, Savonarola saw his opportunity and immediately began to elevate his malicious exhortations. He succeeded in inciting the populace to rise up and depose Piero, and to install himself as the new strongman of the republic. As the spiritual leader of the new political party, the "Weepers," he began instituting a series of diabolical measures that grew until they reached a crescendo in one especially odious incident. In the Bonfire of the Vanities his crazed followers burned countless priceless artifacts—including books and works of art. It is said that Botticelli, swept up by the madness, personally participated in the rampage, flinging some of his own paintings into the fire.

As Savonarola's revolution escalated, so did people's fears. The madness did not end until 1498, when Pope Alexander VI, perceiving the threat to the church, entered the fray. He excommunicated the friar. Stripped of his moral authority, Savonarola was unable to fend off the wrath of the citizens, who on May 23, 1498, hanged him in the Piazza de la Signoria and then burned his body. Had Leonardo been in Florence at the time, no doubt he would have chronicled the event, sketching and recording his observations. Ironically, he may still have acknowledged the evils of Savonarola from a distance—by modeling his Judas in "The Last Supper" after the infamous friar.

Another conjecture, no less tenable, no less befitting of Leonardo lore, is that his Judas was located in the most unlikely of all places. In this account, Leonardo had selected and even painted his Jesus along with eleven of the twelve disciples, but had been stymied in locating this last and most reprehensible character. But when a close friend visited the artist and

told him of having met a bedraggled wretch of a man—an inmate in a local prison—Leonardo pursued the lead, and visited the man. After the prisoner agreed to serve as the model for Judas, and Leonardo began drawing him, the man remarked, "You don't recognize me, do you!" Leonardo looked up, and suddenly began to tremble. This was the same man he had used for his Jesus a year or two earlier. Having subsequently fallen on hard times, he had run afoul of the law and been incarcerated.

Above: A 16th-century painting depicting Savonarola being burned at the stake in the Piazza della Signoria in 1498. He had been charged with heresy, sedition, and religious error.

Opposite: Monument to Savonarola, in his hometown of Ferrara. Often cited as a precursor of Martin Luther and the Protestant Reformation, the Dominican priest preached against moral corruption in the church.

Federico Barocci, "Last Supper," 1599. Chiesa della Minerva, Urbino. A proto-baroque artist, Barocci was born Federico Fiori, but preferred to go by his nickname, "Il Baroccio," literally a two-wheeled oxen cart. Although this rendition of the Last Supper was painted after Leonardo's mural, the Judas is positioned with his back to the viewer, the traditional, accepted pose that Leonardo eschewed for his depiction of the event.

In the mode of a modern film director seeking just the right actors for his cast, Leonardo spent months observing, sketching, and selecting from the storehouse of faces in his notebooks, in search of just the right countenance for each disciple. He sought expressive faces for all, but finding the models for Christ and Judas was especially critical. After a while he found the model for Christ, a handsome young man exuding confidence and an air of piety. Nonetheless, Christ's face appears a bit nebulous—whether this ambiguity was Leonardo's original intention, or whether the face has been painted over by lesser artists leaving us with the present visage, we cannot tell. The casting for Judas was still some time away.

The Destruction of "The Last Supper"

Almost immediately after its unveiling, "The Last Supper" was recognized for its incomparable mastery—its thunderclap power and sublime beauty—inspiring praise and myriad copies. Just a year after its completion, when the French armies of Louis XII conquered Milan, the king expressed an interest in having the entire brick wall on which the mural is painted transported en masse to France, but was dissuaded by his engineers' arguments that any such attempt would bring about the total destruction of the mural. Just a few years later, his successor, François I, and three centuries later, Napoleon, revisited the notion of exporting the wall to France. Each time the court architects and engineers offered the same argument as that of the earlier engineers. Each time the monarchs relented, and left the work in place. But in 1796 French troops, who used the refectory as a barracks, vandalized it, throwing stones and climbing ladders to scratch out the eyes of the subjects—clear evidence that genius may have its limitations, but stupidity has none. On April 15, 1943, during some of the heaviest fighting of World War II, a bomb dropped from an Allied airplane landed on the church, sparing Bramante's dome, but knocking down a wall of the refectory at a right angle to the one occupied by the mural. Miraculously the wall and the mural survived, thanks first to the fact that it was not a direct hit on that wall, and second to sandbags stacked against the wall in order to stabilize it.

Yet a painful irony should not be forgotten. The great mural had already been doomed to destruction by its creator's own experimental techniques. The time-tested fresco method for a mural called for the application of pigments dissolved in water to be applied directly onto freshly laid plaster, assuring the colors' penetration of the plaster to a depth of up to a millimeter. The method, known as *buon fresco* (true fresco), was the classic technique of artists producing murals from the late 13th to the mid-16th century. A discouraging aspect of the method was that the painter was forced to work quickly; the salutary effect was the resilience of the work to the ravages of time as the pigments and the plaster became integrated. Giotto's works have survived for more than 700 years; the frescoes of Pompeian artists, similar to those of medieval artists, have survived 2,000 years. Leonardo, with his need for lengthy deliberation and frequent changes in the design, needed a medium that would allow him much more time to work. Accordingly, he decided to treat the mural as if it were a wood panel. He applied two ground layers of gesso as primer, one of lead base, directly onto dried plaster. As for the paints, he decided on mixing media, using both oil-based and tempera-based colors simultaneously neither of which would penetrate the primer, and certainly not the underlying plaster. To compound the exigencies were the unusually dry climatic conditions prevailing in northern Italy during Leonardo's endeavors on the mural in the late 1490s. Modern analysis of tree-ring growth in northern Italy from that time reveals the prevalence of widespread drought. This would have expedited the drying of the primer, bringing about early crackling.

Normally in painting a fresco, water-based paints are applied to wet plaster, penetrating and becoming a part of the plaster. Leonardo used oil-based paint along with varnish, and these, applied to a dry wall, never achieved sufficient penetration. Unproven techniques and materials, coupled with the salts leached from

"[O]ne cannot look for long at the "Last Supper" without ceasing to study it as a composition, and beginning to speak of it as a drama."

—KENNETH CLARK

LEONARDO DA VINCI, 1939

THE LAST RESTORATION

"You have to think like the artist.... What you see of the painting is largely the invention of past restorers.... What we are bringing to light will be truly by Leonardo's hand." —PININ

BRAMBILLA BARCILON

NATIONAL GEOGRAPHIC, 1983

Above: Santa Maria delle Grazie after Allied bombing in World War II. Leonardo's "Last Supper" is just behind the hanging tarp.

Opposite: The dome of Santa Maria delle Grazie

By the mid-20th century, Leonardo's "Last Supper" was showing the ravages of time and war: Its colors were faded and obscured by grime and several areas had sustained damage. Between 1951 and 1954, art conservator Mauro Pellicioli carried out a legitimate conservation project, cleaning and stabilizing the mural until a future generation with superior technology could attempt a more comprehensive restoration. The most recent effort to restore the painting—a project that ran from 1978 to 1999 (five times longer than the original creation of the work)—may have finally succeeded in decelerating the mural's disintegration and forestalling its complete and irretrievable loss. Under Dr. Pinin Brambilla Barcilon, the chief conservator of the restoration project, a heroic program was undertaken using the best technological solutions available and those developed precisely for this task. Bram-

billa's techniques reflect those of the modern conservator: Not to impose on the work what she as an artist might have done—as had been the approach of earlier restorers—but rather to get into the mind of Leonardo. Working painstakingly in minute areas, frequently using stereoscopic microscopes, she applied specially created solvents and blotted them off quickly before they penetrated the layers of Leonardo's original colors. To those areas where Leonardo's own pigment had been irrecoverably lost, and painted over by past restoration schemes, Brambilla applied neutral, easy-to-remove watercolors.

The November 1983 issue of NATIONAL GEOGRAPHIC magazine reported the progress of the restoration five years into the project. In 1999 a definitive volume was published, under the authorship of Brambilla, and Pietro Marani, her co-director, chronicling the project's history.

the upwelling groundwater, caused the paint to gradually flake off. Indeed, Leonardo, just two years after completing the project and making a lengthy sojourn away from Milan, returned to find the painting already beginning to perish, the paint to peel and flake. By 1556, according to Vasari, the mural had deteriorated to a "muddle of blots." Aldous Huxley in the early 20th century commented about its appalling state of neglect as "the saddest work of art in the world," and Henry James would characterize it as "an illustrious invalid."

Over the centuries a number of attempts were made to restore the great mural, most of them expediting its decay. The most recent effort to restore it, a project that lasted many years, may have finally succeeded in decelerating further degradation by halting the leeching of salts into the wall. And, in removing the overpainting of restorers, the project brought to light Leonardo's original vivid palette.

Among the legion of artists who have tried to copy or produce variations of Leonardo's "Last Supper" was Raphael Morghen, who produced a superior engraving based on the work (ca 1800) that would help in disseminating the image to the world. It is this engraving that reveals some of the details of Leonardo's symbolism. Among the group immediately to the left of Christ is Judas, face in shadow, clutching a sack of silver in one hand. In one of Leonardo's studies for "The Last Supper" Judas is not sporting a beard, suggesting the facial hair may have been overpainted by a later artist.

On the Road Again

In early 1498 Ludovico was once more involved in trying to readjust the political balance among the Italian powers. Again he encountered more than he could handle. The new King of France, Louis XII, was a descendant of the Visconti, former rulers of Milan overthrown by the Sforza. He announced that he intended to assert his claim to the duchy of Milan. Louis XII allied himself with Venice and bribed Florence to stay out of the conflict. Ludovico was doomed. He had no friends and his coffers were empty from the war of 1495. In fact, he was so low on cash that the house and vineyard just outside the walls of the city that Ludovico had given to Leonardo six months before this time may have been in lieu of cash that was owed.

In May 1499 French forces entered Italy. By July they were approaching the outer territories of the duchy of Milan. On September 14 the city fell without a shot fired, Ludovico and the elite having fled. On October 6 Louis XII entered the city in triumph. Leonardo initially remained, evidently hoping to curry favor with Milan's new masters. By now he was known internationally. Few leaders would not have relished the chance to have such a man as an adornment of their court. And the French greatly admired his work. If Leonardo could have remained in Milan with the same deal from the new regime that he had from the old, he might have stayed. But Louis returned to France and the French troops began to mistreat the populace. They even destroyed the clay model of Leonardo's great horse; Gascon archers used it for target practice. In December there were rumors that Ludovico was about to return with the help of Maximilian. After staying and cooperating with the French, even for a few months, Leonardo was not destined for a happy reunion with Il Moro. On December 14 he transferred his savings to a bank in Florence and prepared to depart with Salai and his friend Luca Pacioli.

Opposite: During the two decades that "The Last Supper" was undergoing restoration, visitors continued to flock to catch a glimpse, obscured though it was by scaffolding

LEONARDO'S SKETCHBOOK

WHILE LEONARDO WORKED FOR THE SFORZA COURT, his engineering projects involved the design of commercial and industrial machines, along with military weapons. His employment with Cesare Borgia's forces would call for more topographic and cartographic work, although he would conceive some strategic weapons. In his engineering designs, we see his extraordinary prescience, prefiguring weapons of the future, as well as the irony of a devoted pacifist being engaged in weapon design.

"System of Defense with Mortars Firing over Castle Walls," ca 1504. Codex Atlanticus, Biblioteca Ambrosiana, Milan

"Drawing of Shields and Cannonballs Exploding on Contact," ca 1485-1500. École Nationale Supérieure des Beaux-Arts, Paris

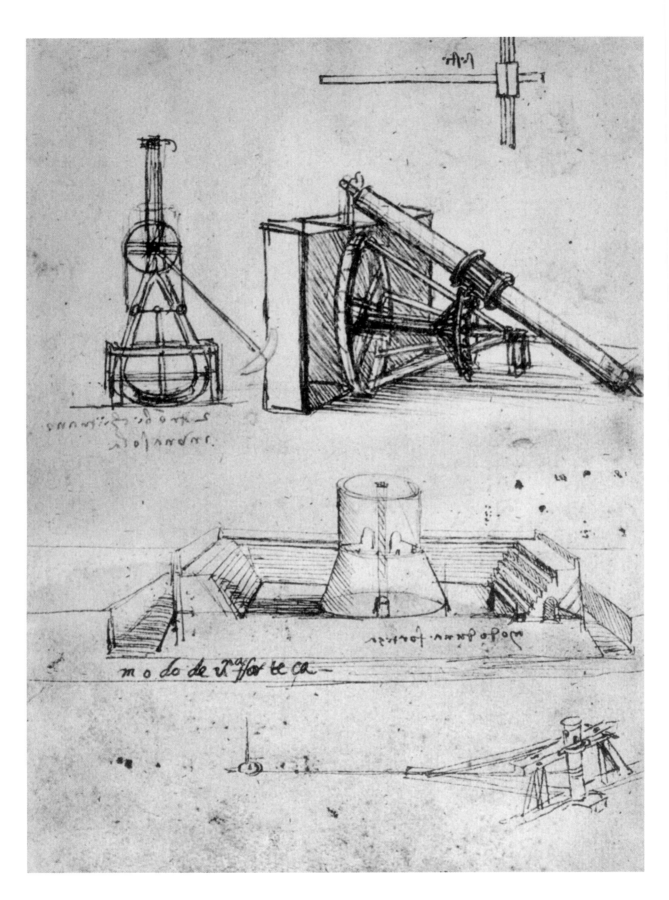

"Drawing of Defensive Weapons," ca 1490-1515. Codex Atlanticus, Biblioteca Ambrosiana, Milan

"Drawing of Apparatus for Scaling Walls," ca 1490–1515. Codex Atlanticus, Biblioteca Ambrosiana, Milan

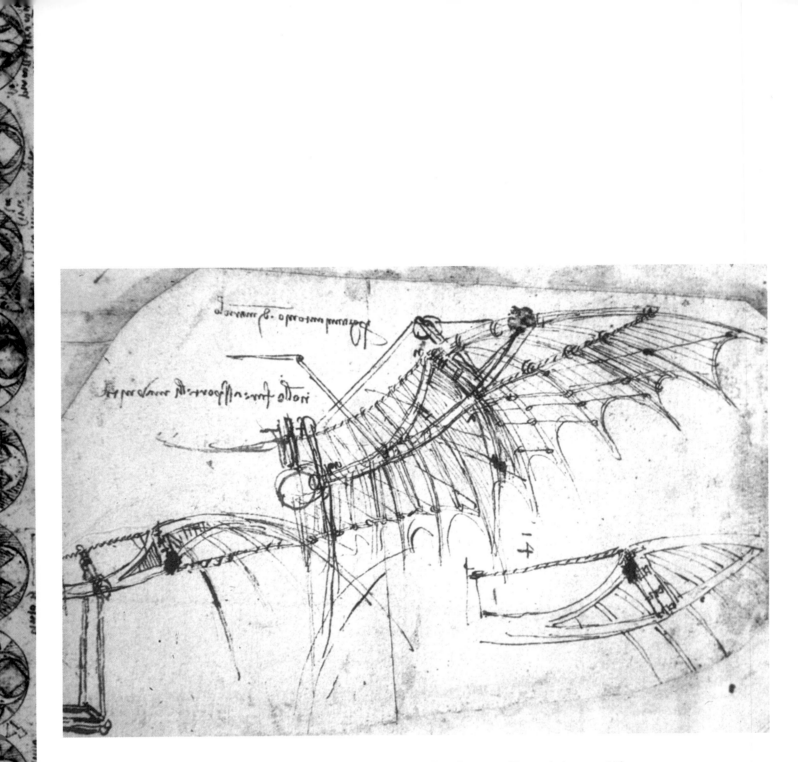

"Glider or Ornithopter Wings," ca 1490-1515. Codex Atlanticus, Biblioteca Ambrosiana, Milan

"Design of a Hydraulic Machine," ca 1480. Paris Ms B, Institut de France, Paris

Chapter Five

"Those who fall in love with practice without science are like pilots who board a ship without rudder or compass."

—LEONARDO

1500-1507
ON THE ROAD

Leonardo's first stop after leaving Milan was Mantua, where he and his entourage were welcomed as guests of the Marchioness of Mantua, Isabella d'Este. Well educated and a great patron of the arts, Isabella was the sister of Beatrice d'Este, who had married Ludovico Sforza in 1491. The Este of Ferrara were among the most powerful and prominent families of Renaissance Italy. Both sisters made politically and financially advantageous marriages—Beatrice to Ludovico of Milan and Isabella to the Marquis of Mantua, Francesco Gonzaga II. Their brother married Ludovico's niece, Anna—a tidy arrangement that joined great fortunes and cemented alliances.

Mantua and Venice

Though Isabella chafed under the knowledge that her sister, Beatrice, had married a somewhat wealthier man, she spent more lavishly on her art collection than almost anyone else in Italy. For her, collecting was an obsession. She even "requested" of Cecilia Gallerani that she surrender her now famous portrait done by Leonardo. She collected paintings, sculptures, coins, miniatures, watches, and all sorts of bric-a-brac and ephemera, much of it done by the most illustrious artists and craftsmen of the time. For Leonardo, she might reasonably have seemed like the ideal choice as new employer and patroness after the loss of Ludovico. She eagerly sought Leonardo's services, asking him to paint her portrait. He did a drawing of her in profile, presumably in preparation for a painting, but there the relationship ended. He did not trust her, and after only a few months in Mantua, he was on the road again, headed for Venice. Over the next few years Isabella begged him by letter and through emissaries to return to Mantua and finish her portrait, but Leonardo answered none of her letters and remained deaf to all entreaties.

As a patroness Isabella had a reputation for interfering with commissions in progress and treating artists badly when she did not get what she wanted. She had taken Giovanni Bellini to court to force him to complete a painting in the manner

Opposite: Gondoliers ferry tourists around Venice's Grand Canal, reflecting a form of transportation that has its roots in the 11th century. Previous pages: Fairy green light filters through a grove of poplars near Lago Inferiore.

she desired. This alone would have been enough to scare off Leonardo. But Isabella also represented a passing political reality. Italy was in turmoil. The French and their allies were triumphant, the new power to be reckoned with. The Sforza and their allies were yesterday's story, and by now Leonardo knew the importance of keeping a finger in the political wind. Venice, his next destination, was a longtime enemy of Milan and the Sforza. Probably, too, he was growing tired of the sort of commissions Isabella offered. Painting was not his main priority. One of Isabella's emissaries to Leonardo reported to her that he "cannot abide the paintbrush." Indeed, he left a trail of promises, abandoned commissions, and few finished paintings from this time.

When he departed Mantua for Venice in the first months of 1500, Leonardo was in the company of Luca Pacioli—the embodiment of Leonardo's interest and fascination with mathematics and engineering. Venice at the time was under military threat from the Turks, who had landed an army just two days' march from the city. Leonardo had likely been invited to Venice as an engineer to recommend defensive measures to be used against the invaders.

Within weeks of his arrival there, he was consulting with the governing council on measures to be taken. Leonardo decided that a dam should be built across the Isonzo River, north of the city, behind which a large body of water could be collected and then let loose should the Turks decide to move along the river valley toward the city. The idea was rejected. He next proposed the use of divers—using breathing equipment of his design by means of a tube that ran from a diving mask to the surface—to attack and sink the Turkish fleet while at anchor, trapping the Turks in Italy. The authorities thought the enemy might spot the tubes floating on the surface, so Leonardo designed a diving suit that carried its own air supply in wineskins. This scheme too went nowhere, and he finally left Venice.

Return to Florence

In April 1500, Leonardo's wanderings took him back to Florence after an absence of 18 years. In some ways he returned as a conquering hero. Leonardo Fiorentino—Leonardo the Florentine—was now famous all over Italy and beyond. He was, after all, the creator of the great Sforza horse and the incomparable "Last Supper." His pageants with their wondrous machinery were spoken of as far away as France. He was immediately granted a commission to paint an altarpiece, the "Virgin and Child with St. Anne," given originally by the Servite Friars of the church of the Santissima Annunziata to Fra Filippo Lippi, who obligingly withdrew in favor of Leonardo. Vasari tells us that the commission included lodgings in the Annunziata and expenses for Leonardo and his entire household. His welcome back to the city could hardly have been warmer, but there was also a new kid on the block, Michelangelo Buonarroti. They met—and took an immediate dislike to each other.

Leonardo's work on the "Virgin and Child with St. Anne" proceeded slowly. Sometime in 1501 he completed a full-size cartoon of the proposed painting. The unveiling of the cartoon at Santissima Annunziata was the artistic event of the year in Florence. Vasari says that for two days the room where it was exhibited was crowded with men and women who looked at the work in amazement.

Above: Leonardo, "Drawing of Isabella d'Este," ca 1500. Musée du Louvre, Paris

Opposite: Leonardo, "Map of the City of Imola," 1502. Pen and ink and watercolor. Royal Collection, Windsor Castle, U.K.

Subsequently, the typical Leonardesque pattern asserted itself. Progress on the work came to a halt. Another cartoon (known today as the "Burlington House Cartoon") was made as late as 1508, and the painting was not finished until 1510.

Leonardo's other interests—mathematics and geometry—were taking up much of his time. Luca Pacioli, his friend and mathematics mentor, soon joined Leonardo in Florence. In early 1501 Leonardo also made his first visit to Rome. The visit was brief, but he must have met with his old friend Bramante, who would soon be at work on the new design for St. Peter's Basilica. Leonardo was restless throughout the remainder of 1501. He put aside the commission for the Servite

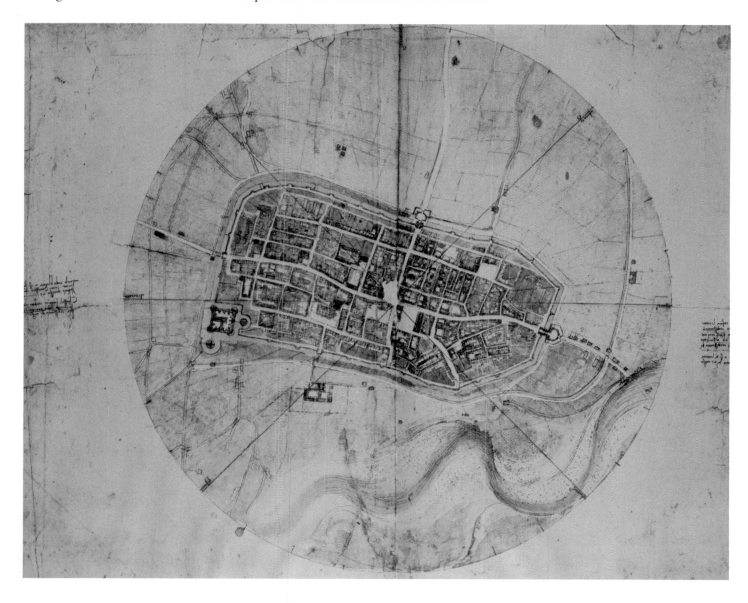

Friars. When he painted, he worked on the "Madonna of the Yarnwinder," a commission initiated by King Louis XII of France for one of his courtiers—perhaps careful political calculation by Leonardo. With France then ascendant in Italy, Leonardo may have been looking out for his future.

In the summer of 1502, Leonardo departed Florence to act as a military engineer for Cesare Borgia. Exactly how he came to be in Cesare's employ is not known.

Cesare was the illegitimate son of Pope Alexander VI, and like his father he was ruthless, relentless, and thoroughly unprincipled. He had allied himself to the French when they invaded Italy in 1499. Married to Louis's cousin, Charlotte d'Albret, he had entered the city of Milan with Louis XII in 1499, and Leonardo probably first met him at that time. Leonardo's job consisted mostly of inspecting the fortifications of cities and towns that Cesare had conquered, insuring that they were ready to repel any attackers. Cesare gave Leonardo a passport that commanded all military and civilian authorities to render Leonardo any service or assistance he might require and to follow his instructions in all things related to his responsibilities.

"Men ought either to be indulged or utterly destroyed, for if you merely offend them they take vengeance, but if you injure them greatly they are unable to retaliate, so that the injury done to a man ought be such that vengeance cannot be feared."

—MACHIAVELLI

Leonardo traveled around Cesare's extensive realm in central Italy and surely enjoyed seeing the countryside as well as new towns and cities. From these travels came the observations necessary for many of Leonardo's topographical maps, more detailed and accurate than any such maps up to that time. When Cesare temporarily established his court in the city of Imola, Leonardo joined him and prepared his famous map of that city and its surrounding fields. And in October 1502, he met the young Florentine envoy to Cesare, Niccolò Machiavelli (1469-1527), who later authored the most important political tract of the Italian Renaissance, *The Prince*. Machiavelli's model for the political behavior described in *The Prince* was Cesare Borgia, whose single-minded and sure-footed instinct for survival and unsentimental exercise of power in the unforgiving world of Renaissance Italy thrilled and amazed the young ambassador. Leonardo and Machiavelli became friends and mutual admirers.

In the autumn of 1502 rebels staged an attempted coup against the rule of Cesare Borgia. His response was stern and decisive. In December, one opponent, Ramiro de Lorqua, is sliced in two and left lying in the piazza in the city of Cesena. Leonardo recorded the sight in a notebook. Cesare invited the ringleaders to a meeting billed as an opportunity for reconciliation, but when they arrived he unceremoniously seized and bound the rebels. Machiavelli grimly reported to his superiors: "[T]hey will not be alive tomorrow morning." Most were, in fact, strangled that night. All were dispatched within a matter of weeks. Machiavelli took such events in stride. Leonardo may have had qualms about working for such a man, or even uneasiness concerning his own safety.

By March 1503, Leonardo was back in Florence—and trying to get released from the commission to complete the "Virgin and Child with St. Anne." The friars, who had grown impatient, were undoubtedly retracting their generous offer of support for the Leonardo entourage. Forced to rely on his savings, he began to make regular withdrawals from the bank. Fortunately a new project—similar to the one the Venetians would not risk—presented itself.

During the French invasion of Italy in 1494, the Florentine-controlled city of Pisa had been given to the French in exchange for a French pledge not to attack Florence. When the French withdrew, Pisa declared independence, and Florence subsequently had been trying to take the city back. In the summer of 1503, possibly on the recommendation of Machiavelli, Leonardo became chief engineering consultant on a project to divert the Arno River so that it would bypass the city of Pisa, thereby cutting Pisa off from its vital access to the sea. The plan called for the river to be dammed and the flow diverted into two 12-mile-long canals that would drain into a marsh known as the Stegno. Leonardo calculated how many man-days would be required

Above: Santi di Tito, "Portrait of Niccolò Machiavelli," ca 1572. Palazzo Vecchio, Florence. Santi di Tito painted this posthumous portrait of the statesman some 45 years after his death.

Opposite: Leonardo, "Studies of Flowing Water," 1508-1510. Pen and ink over red chalk. Royal Collection, Windsor Castle, U.K.

A BRIDGE FOR THE SULTAN

A letter was found in the archives of the Topkapi Museum in Istanbul in 1952. It is in Old Turkish (ironically written from right to left, similar to Leonardo's Italian mirror script) and, according to its heading, is a translation of a letter addressed to Sultan Beyazid II "that the *gâvur* [a derogatory expression meaning "infidel" in Turkish] named Lionardo sent from Genoa." The original letter was probably written in 1503. In the letter Leonardo offers his services as architect and engineer to design and build a bridge across Haliç (the Golden Horn). The Golden Horn is a narrow estuary that divides Istanbul and flows into the Bosporus. Early that year Leonardo had been in Rome and was no doubt aware that ambassadors from the sultan had recently been in the city, and that they had announced the sultan's desire to engage an Italian engineer to construct such a bridge over the Golden Horn. Vasari says that Michelangelo turned down a direct request to lead the project.

Leonardo was always on the lookout for the main chance, and with the unstable political situation in Italy and the volatile and violent reputation of his then employer, Cesare Borgia, a well-paid sojourn to the Ottoman capital would have looked attractive. So he sent the letter. And just as he had done in the earlier letter to Ludovico Sforza, he started with boasts about the engineering wonders he could perform—innovative windmills and water pumps and such; however, this time he wasted little space on these preliminaries before getting down to business. The most astonishing passage reads, "I understand that you have expressed an interest in building a bridge from Istanbul to Galata, but you have not proceeded, having failed to find a man who knows how to build [the bridge]. I, your obedient servant, know how to build it." He proposes a long, very high drawbridge that will allow the passage of the masts of tall ships. But in a notebook of the same period he sketches a somewhat different design—a tall, streamlined, sweeping structure with bird-tail-like abutments, formed essentially from two parabolic arches leaning against one another, that efficiently distribute the weight of the bridge and support a roadbed that climbs to 150 feet above the water. At 1,200 feet long, it would have been the longest masonry bridge in the world. The design was functional, but also striking and beautiful; however, as far as it is known, the sultan never saw Leonardo's sketch. And, for whatever reason, no bridge was built over the Golden Horn.

Not wishing to see a great design idea languish, Norwegian artist Vebjorn Sand built a pedestrian bridge, based on Leonardo's design, over Highway E18 in Ås, Norway, about 20 miles south of Oslo. The scaled-down 300-foot version of the bridge, made of laminated wood, opened in October 2001, nearly 500 years after it was born in the mind of Leonardo.

Opposite: Leonardo, "Studies for a Proposed Bridge Spanning the Golden Horn," ca 1503. Institut de France, Paris

Above: Vebjorn Sand's Leonardo Bridge, christened in Ås, Norway, in 2001

to dig the canals based on his estimate of how much earth would have to be moved. He reckoned it would take 54,000 man-days to complete the project. He even devised a number of excavating machines that he thought would be more efficient and thus shorten the time required. Work began in August 1504 and proved much more difficult than anticipated. In October a severe storm caused the collapse of the walls of the ditches that had thus far been dug and the loss of the lives of about 80 workmen. The project was then abandoned after an expenditure of more than 7,000 ducats.

"Leda and the Swan"

The story of Leda and the Swan is from Greek mythology. Zeus, a notorious womanizer, transforms himself into a giant swan and seduces Leda, the lovely unsuspecting wife of Tyndareus, King of Sparta. As a result of the union, Leda conceives and lays a pair of eggs, each producing a pair of twins—Castor and Pollux from one egg, and Clytemnestra and the beautiful Helen of Troy from the other. Leonardo's allegorical painting of the story, "Leda and the Swan," was produced in a period that overlapped his endeavors on the "Battle of Anghiari," the "Mona Lisa," and the "Virgin and Child with St. Anne." In his days in Milan, Leonardo had frequently sketched the swans around the moat of the Castello Sforzesco, as well as a number of head and bust studies for Leda. He had also drawn two separate full-size cartoons: in the first, conceived in 1504, Leda is kneeling, writhing, which, according to art historian Maxine Anabell, is suggestive of her fertility; in the other, she is standing, with head slightly bowed, and while appearing to be recoiling from the swan, she displays a strange attraction to the bird. It is the latter drawing (produced in 1506) that served as the cartoon for Leonardo's painting. The demure countenance is strongly reminiscent of St. Anne in "Virgin and Child with St. Anne," as well as the subject of the "Mona Lisa." The standing nude figure of a woman in contrapposto posture was unique for Leonardo.

Leonardo's "Leda" has been lost since the 18th century. It was most likely in the possession of Louis XII of France early in the 16th century. In its time it inspired copies by many painters, including Raphael in 1506, as well as Correggio a little later. Giulio Bugiardini produced a painting based on Leonardo's first cartoon, as did Leonardo's assistant, Francesco Melzi, in 1508. A painting by another assistant, Cesare da Sesto, is in line with Leonardo's second drawing and thought to be a copy of Leonardo's original "Leda." In this work, Leda's face and coiffure bear a strong resemblance to sketches Leonardo left in his notebooks. Maxine Anabell, in her succinct but authoritative account, quotes Cassiano del Pozzo, who claims to have seen Leonardo's painting in the royal collection in Fontainebleau in 1625:

A standing figure of Leda almost entirely naked, with the swan at her [feet] and two eggs, from whose broken shells come forth four babies. This work, although somewhat dry in style, is exquisitely finished, especially in the woman's breast; and for the rest, ... the landscape and the plant life are rendered with the greatest diligence. Unfortunately the picture is in a bad state, because it is done on three long panels which have split apart and broken off a certain amount of paint.

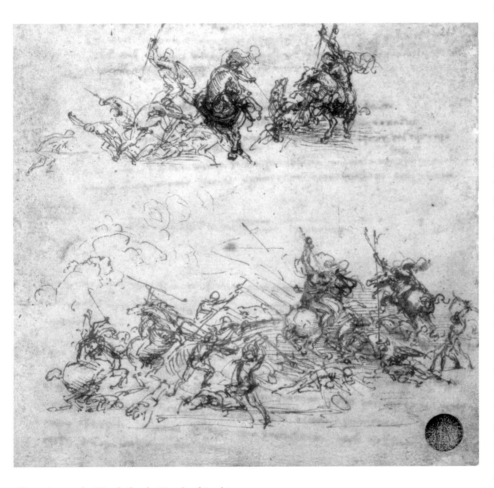

Above: Leonardo, "Study for the 'Battle of Anghiari,'" ca 1503. Galleria dell'Accademia, Venice

Opposite: Michelangelo, "Study for the 'Battle of Cascina,'" ca 1504. Gabinetto dei Disegni e delle Stampe degli Uffizi, Florence

The long-held belief that the painting decorated the steam-saturated Royal bathhouse in Fontainebleau would explain the deteriorated state that del Pozzo described.

Battle of the Titans

Leonardo and Michelangelo, in 1503 and 1504, respectively, were awarded a pair of commissions by the Signoria to paint large adjacent murals on the eastern wall of the Sala del Maggiore Consiglio, or Council Hall, of the Palazzo della Signoria in Florence. (Today, the room and building are known as the Sala del Cinquecento and Palazzo Vecchio, respectively.) Leonardo was to paint the "Battle of Anghiari," celebrating the Florentine victory over the Milanese in 1440, on the northern half of the 177-foot-long wall; Michelangelo, fresh from the completion of his statue "David," was to paint the "Battle of Cascina," celebrating the Florentine victory over the Pisans in 1365, on the southern half. The murals were to dwarf "The Last Supper." The "Battle of Anghiari" alone would have measured 57.4 feet by 23 feet. The particular commission to Leonardo came from Piero Soderini, the chief government official, and included Niccolò Machiavelli as a signatory to the document.

What should have been a pair of masterpieces created by the titans of the High Renaissance came to an inglorious end. Leonardo had gone so far as to prepare his cartoon, transfer it to the wall, and begin an underpainting. He employed another experimental fresco technique, and the mural appeared headed toward abject failure. The paints began to run despite his attempts to dry the wall by applying heat. Michelangelo, meanwhile, had completed the cartoon for the "Battle of Cascina," but abandoned the project abruptly in 1505 when he was ordered to report to Rome to create the tomb of Pope Julius VI.

Both cartoons hung for several years on the walls designated for the frescoes. Then Michelangelo's was destroyed in 1512 by a jealous artist, Bartolomeo Bandinelli, who cut it up into pieces. It is not known how long Leonardo's cartoon survived, or whether it even lasted as long as Michelangelo's. But a copy of the cartoon for the "Battle of Anghiari" was produced, albeit by a poor artist, and this was still in existence a century after Leonardo first conceived the piece. In 1603 Peter Paul Rubens, the Flemish baroque master, inspired by this copy, produced a drawing in the "style of Leonardo" that also incorporated Rubens's own style of fleshy creatures—especially the horses. Considering Rubens's skill

as an artist, and his reverence for Leonardo, it is likely that he rendered as faithfully as possible what he thought Leonardo had in mind; he certainly transcended the quality of the copy used in creating his drawing. The drawing echoes Leonardo's personal aversion to the horrors of combat, notwithstanding the irony of his employment as a military engineer. Human and equine warriors are seen caught up in a virtual maelstrom, all intertwined, growling, thrashing, biting. Some of Leonardo's sketches of battling warriors and horses can be found in his notebooks; the faces are highly reminiscent of those of some of the warriors and horses in Rubens's version of the drawing. In its pyramidal design, it is signature Leonardo.

UNCOVERING THE COVER-UP

"Make the conquered and the beaten pale, with brows raised and knit, and the skin above their brows furrowed with the pain... the lips arched, displaying the upper teeth, and the teeth apart as with the crying of lamentation."

—LEONARDO

Maurizio Seracini, a latter-day compatriot of Leonardo, cobbles together expertise in different fields. The Florentine, trained in bioengineering at the University of Padua and in art history at the University of California–San Diego, is the founder of Editech, an art diagnostics company that has garnered an impressive record of success in answering questions regarding techniques of old artists and the provenance of great paintings, as well as uncovering hidden paintings. The revelation that Leonardo's unfinished "Adoration of the Magi," hanging in the Uffizi Gallery, had been contaminated by the hands of lesser talents, perhaps as much as a century after Leonardo had abandoned it circa 1482, came from a four-year study carried out by Seracini's laboratory; so did the original positions of the Three Graces in Botticelli's masterpiece "Primavera" (or "Allegory of Spring"), also in the Uffizi—hanging in the Botticelli Room, located next to the Leonardo Room.

Along with traditional archival techniques, Seracini invokes 21st-century technology—from scanning with ultrasound probes to scanning with electromagnetic radiation in a wide spectrum of ranges, from the ultrashort x-radiation to ultralong infrared. In the Palazzo Vecchio project to uncover the lost "Battle of Anghiari," he is employing neutron activation analysis. This is a technique normally used at nuclear reactors, where neutrons emanating from the core of a reactor are employed in activating samples under investigation (specifically the nuclei of the chemicals present in the sample). Subsequently, gamma-ray analyses of the activated specimens reveal the culprit chemicals. Thus neutron activation is sometimes used as a medical diagnostic tool, capable of detecting impurities of one part in a trillion, or tantamount to a needle in a haystack the size of the Great Pyramid. Hair and fingernails become repositories of heavy metals, and trace amounts of such metals can be detected in the process. For example, we knew that Isaac Newton suffered a catastrophic mental breakdown in 1693, but not the cause, until a neutron activation analysis performed on four strands of his hair determined that Newton had accidentally poisoned himself with the mercury he was handling in his alchemical studies.

Neutrons, as neutral particles, can penetrate walls with ease and activate chemicals at different depths. Since Leonardo left meticulous notes regarding the ingredients he used in his paints and primers—for example, vermillion (which contains mercury), the pure whites (which almost always contain lead)—by scanning the Vasari wall with a beam of neutrons, it is hoped the presence of the lost Leonardo mural will be revealed. In November 2007, Seracini and his 30-man crew were engaged high on a scaffolding, using a specially scaled-down apparatus for the neutron source to scan the wall, and gamma-ray detectors to identify the ingredients. The analysis will take some time. And the art world will hold its collective breath, hoping that the lost masterpiece will reveal itself.

Above: Dr. Maurizio Seracini stands below Vasari's mural, believed to have been painted over Leonardo's unfinished "Battle of Anghiari."

Opposite: A colossal mural by Giorgio Vasari and his assistants adorns the wall in the Palazzo Vecchio originally intended for the complementary murals by Michelangelo and Leonardo.

"You will make the conquerors rushing onwards with their hair and other light things streaming in the wind, with brows bent down."

—LEONARDO

Giorgio Vasari, detail from "Battle of Marciano in Val di Chiana," 1563. Sala del Cinquecento, Palazzo Vecchio, Florence

In 1563 Giorgio Vasari was hired to remodel the Sala del Maggiore Consiglio and to paint over the incomplete work left by Leonardo. Vasari in his *Lives* extolled Leonardo's composition: "It would be impossible to express the inventiveness of Leonardo's design ... the incredible skill he demonstrated in the shape and features of the horses, which Leonardo, better than any other master, created with their boldness, muscles and graceful beauty." Vasari's cover-up—with his own colossal mural, "Battle of Marciano in Val di Chiana"—was all too successful, defying subsequent attempts to locate the unfinished mural. There is a possibility, however, that Vasari, because of his

Peter Paul Rubens, "Battle of Anghiari," also known as the "Battle for the Standard," 1603. Musée du Louvre, Paris. This drawing is believed to be a copy of the center section of Leonardo's "Battle of Anghiari."

immense admiration for Leonardo, might just have built a false wall in front of the original Leonardo–Michelangelo wall, before proceeding to create his own painting. In the clamorous scene of his "Battle of Marciano" there is a small banner carried by a soldier, too high and too small to be read from the floor. It reads: "*Cerca Trova*—He who seeks, finds." Inspired by his mentor, legendary Leonardo scholar Carlo Pedretti, art diagnostician Maurizio Seracini has devoted three decades to developing the modern technology necessary to become that ultimate seeker.

"Mona Lisa"

In 1503 Leonardo accepted a commission by a wealthy Florentine silk merchant, Francesco del Giocondo, to paint a portrait of his young wife, the former Lisa di Antonmaria Gherardini. Vasari was the first to refer to the subject of the painting as Mona Lisa, and it is the Italian appellation Madonna Lisa, in its contracted form, that gives us Mona Lisa. In Italy the portrait is known as "La Gioconda," in France as "La Joconde," presumably referring to the subject's married name. But the lady's beguiling, enigmatic smile suggests that the reference might be to the Italian adjective, *giocondo,* meaning "jocular"—and in the feminine form, *gioconda,* "a light-hearted woman." Those who have doubted Vasari's identification of the subject have claimed that the title works well as a description of the portrait and need not necessarily refer to Lisa del Gioconco. But the double entendre is the type of play on words that Leonardo would have liked, and which had also occurred in the portraits of the other two "girls"—Ginevra and Cecilia. Early in 2004, Italian schoolteacher and art devotee Giuseppe Palanti announced, after 25 years of archival research, that Vasari's original identification of the sitter had been entirely on the mark. Lisa Gherardini was born in Chianti, married Francesco del Giocondo in 1495 (which made her about 25 in the painting), bore five children, and outlived her husband. The portrait was produced on the occasion of the delivery of her second son.

During the five years that passed between completion of the "The Last Supper" and the beginning of the "Mona Lisa," Leonardo had undertaken very few new paintings, and those he had begun remained unfinished. When he painted the "Ginevra de' Benci," he had been a young artist trying to get established, and Verrocchio may have passed on the commission to him. With the "Portrait of Cecilia Gallerani," the mistress of Ludovico Sforza, he had probably just wanted to impress his patron. With the portrait of the wife of del Giocondo, why he accepted the commission is a mystery. We can speculate that perhaps he needed money—he was making regular withdrawals from his savings at the time—or that he was doing a favor for his father, Ser Piero, who did work for del Giocondo. But one thing is certain: With the "Mona Lisa" he produced a miraculous psychological portrait—spellbinding, hypnotic, timeless. You know this woman, and yet you do not. She is looking directly at the viewer, but what is on her mind is the real enigma. She exudes confidence and uncertainty at the same time, with an expression that is both inviting and frightening. The painting is a universal and enduring statement about the feminine mystique—by a man whose attitude toward the opposite sex was highly problematic, highly ambiguous. In the late 19th century, Renaissance scholar and art critic Walter Pater called the timelessness with which Leonardo had imbued the subject of his portrait, "older than the rocks among which she sits."

Most great art is open-ended, with different viewers taking away different impressions. It is no wonder that Mona Lisa's countenance, with the partial smile, has launched more wild speculation than any other work of art. Among the theories: She is pregnant; she is suffering from a toothache; it's a self-portrait by Leonardo. And the speculation has not subsided with the passing centuries. Eighty percent of the visitors to the greatest art gallery in the

CHOOSING A SPOT FOR THE "DAVID"

Early in 1504, as Michelangelo's magnificent sculpture of David neared completion, the works department of the Duomo in Florence put together a special committee to decide on the best location for the permanent display of the statue. Its membership comprised many of the most prestigious artists then resident in Florence—Sandro Botticelli, Lorenzo di Credi, Pietro Perugino, Filippo Lippi, among many others, including Leonardo.

Leonardo thought that the best place for his rival's masterpiece would be in the Loggia dei Lanzi, "behind the low wall where the soldiers line up," where it would "not interfere with the ceremonies of state." Only one other member of the committee agreed with Leonardo. His obvious scorn for the work of this new upstart, especially in the face of the general recognition of Michelangelo's genius, speaks of the depth of the antagonism that developed between the two artists.

The committee instead gave the statue an honored place in the Piazza della Signoria that it occupied for centuries, and that is occupied today by a replica, the original having been moved into the Galleria dell'Accademia in the 19th century.

world, the Musée du Louvre in Paris, stroll through the museum freely admitting their primary motivation for the visit is to view the most famous painting in the world.

After each 15-year hiatus from painting portraits of women, Leonardo returned with immensely greater knowledge and insight, and a more refined technique. This is not difficult to understand. It is, in fact, quintessential Leonardo. Leonardo accepted the commission for the "Mona Lisa" in 1503 but did not complete it until 1507, although some will argue, including Vasari, that he never finished the portrait—that he simply abandoned it. For whatever

reason—either because he had not completed it in a specified time, or because he had become too fond of it and could not part with it—he never turned it over to Francesco del Giocondo.

Leonardo left Florence with the painting in 1508, and kept it with him through his subsequent travels. He took it with him in 1516 when he moved to France, and it was still in his possession at his death in Amboise in 1519. It was in Amboise that François I purchased the "Mona Lisa" for his local château. It eventually found its way to the royal château at Fontainebleau, outside Paris. It was subsequently moved to the Palace of Versailles by Louis XIV. After the French Revolution, the painting found a home in the Louvre, also the home of Napoleon Bonaparte; by some accounts, the "Mona Lisa" may have decorated Napoleon's field tent during his eastern campaigns. Later, when Napoleon was sent into exile, the "Mona Lisa" returned to the Louvre, which has since been its permanent home.

Above: Leonardo, red ink drawing of a landscape comprising hills, rocky peaks, and a meandering creek, possibly in the Valdarno in Arezzo, 1503. Royal Collection, Windsor Castle, U.K.

Opposite: Leonardo, "Mona Lisa," 1503-07. Musée du Louvre, Paris

THE HEIST OF THE "MONA LISA"

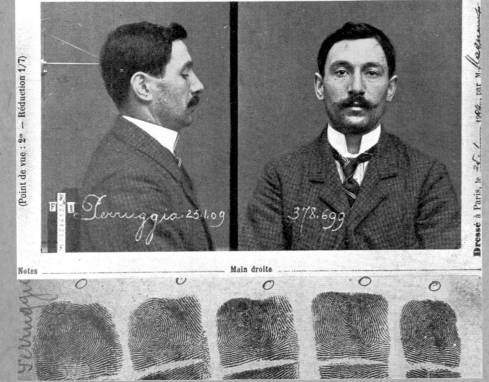

Taille 1ᵐ 61.4		longᵣ 17.7	Pied g. 25.3		nº de cl. 54	Cheveux	Age appᵗ . Age déclaré . Né en 18
Voûte	Tête	largᵣ 15.1	Médius g. 11.0		aurlᵉ	Barbe	
Enverg. 1ᵐ 61		zygᵣ 133	Auricᵣ g. 8.5		périᵣ	Teint Pᵒⁿ	
Buste 0ᵐ 87.8	Oreille dr. 6.6		Coudée g. 43.1		partés	Main dr.	
						Main g.	

Notes   Main droite

Terruggia 25.1.09   378.699

In its 500-year history, the "Mona Lisa" has survived appalling ordeals. It traveled widely in the possession of its creator, and even purportedly decorated Napoleon's field tents during the marshall's eastern campaigns. During the Franco-Prussian War (1870-71), the painting was placed in hiding. During World War II it was hidden again—shipped first to the Château d'Amboise, then to Loc-Dieu Abbey. In 1956 a madman threw acid on the painting, substantially damaging the panel's lower area. And again in the 1960s another madman stepped up and slashed it.

Among the travesties that have visited the "Mona Lisa," the most bizarre took place early in the 20th century. On August 21, 1911, contemporary painter Louis Béroud was visiting the Louvre when he noticed that the space between Correggio's "Mystical Marriage" and Titian's "Allegory of Alfonso d'Avolos," normally occupied by the "Mona Lisa," appeared

prominently empty. A query into the work's whereabouts produced the response from the security officers that it was probably being photographed in the gallery's workshop. When it was determined that the painting was not in the hands of the photographers or conservators, and indeed that it was missing altogether, the Louvre was closed down for a week, with vigorous investigations launched immediately. Initially the police suspected Guillaume Apollinaire, an avant-garde poet who was known to have called for the destruction of the entire museum. Apollinaire was arrested, and his close friend, artist Pablo Picasso, was brought in for questioning as well. Both men were eventually released, and it gradually dawned on the museum authorities that the portrait with the inscrutable smile might be irretrievably lost.

Two years later Vincenzo Peruggia, an Italian nationalist, was arrested as the audacious thief. An employee of the Louvre, Peruggia

had entered the building housing the "Mona Lisa" during regular hours, then hid in a broom closet as the museum was closing down at the end of day. Later he emerged from the closet and took the painting from the four iron pegs attached to the wall. Concealing the painting under his trench coat, he casually strolled across a courtyard and exited the Louvre. He had been in collusion with a con artist named Eduardo de Valfierno, who masterminded the heist and who planned to have the painting copied by a skilled forger, Yves Chaudron, and copies sold to unsuspecting and unscrupulous private collectors. Peruggia spirited the painting to Florence, where he stored it in his apartment, under his bed. Valfierno had decided that as long as the painting was out of the Louvre, and its location unknown, then there was no need to physically have it in hand. He could just begin selling forged copies as the "original." Left hanging by Valfierno, Peruggia became desperate for money; in December 1913 he tried to sell the "Mona Lisa" to a Florentine art dealer, who realized it was genuine. And there the plot unraveled. After the painting was recovered, it briefly toured Italy, being seen by thousands, before it was returned to its wall in the Louvre.

Above: Mug shots and fingerprints of Vincenzo Peruggia. Peruggia's demeanor, trench coat, and turned-down soft hat inspired the bumbling Inspector Clouseau of the Pink Panther *movies.*

Total Integration of Science and Art

The "Mona Lisa" shares with the "Ginevra de' Benci" and the "Cecilia Galle-rani" some mathematical symmetries and geometric constructions well worth noting. First a vertically formatted rectangle can be constructed, enclosing the area from the top of her head down to the top of her bodice. The aspect ratio—its width-to-height measurements—forms the proportion of 1:1.618. This is the celebrated golden rectangle. A square delineated in the upper portion of the rectangle leaves her chin resting on the bottom edge of the square, and her left, or "leading," eye located at the center of the square. This is also true for the "Ginevra de' Benci" and the "Cecilia Gallerani," although in those paint-ings it was the right eye. Finally, the torso of the "Mona Lisa"—slightly turned, her right shoulder and her right cheek set back relative to the left shoulder and left cheek respectively—can be inscribed in a golden triangle (with angles 72°-36°-72°). Was this all a coincidence for Leonardo—just a manifestation of his painterly eye, as it most likely is in the majority of works through the ages where it appears—or was it a conscious exercise? In the work of any other artist, we would assume these manifestations to be coincidental. For Leonardo, who seam-lessly integrated mathematics, science, and art and spent his life seeking unifying principles, perhaps not.

Brunelleschi had developed the principles of one-point perspective early in the 15th century. Leonardo introduced two-point perspective: In the image of a building, as seen from the corner, the roofline on the two visible sides of the building, the lines of the foundation, the lines of windows can all be extended to meet at a pair of vanishing points on the horizon line. In the "Mona Lisa" there is a gentle helical posture, resembling a modern corkscrew, with her torso turned slightly to the viewer's left, the head turned away from the torso slightly to the viewer's right, her eyes directed straight at the viewer, giving the impres-sion of watching her from "a corner." And therein is a display of two-point per-spective in the portrait. Moreover, there is in the painting's design Leonardo's characteristic pyramidal composition.

Toward the close of the 20th century, British-born psychologist and art aficionado Christopher Tyler made a serendipitous discovery known as the cen-terline principle, which states that in a preponderance of great single-subject portraits a vertical line drawn to bisect the painting passes through (or very close to) one eye of the subject. The eye can be the leading eye or the trailing eye. Because this principle has never been taught in art schools and is generally unknown to painters, it is likely an intuitive sense of aesthetics—the artistic or painterly eye of the individual artist—that is responsible for the consistency with which the principle shows up through the centuries. The discovery was covered prominently in the *New York Times* and made a huge splash in the art community. Tyler's matrix, or array, of nine famous portraits through the ages of course includes the "Mona Lisa." When the test of the centerline principle is applied to the two earlier Leonardo portraits—the "Ginevra de' Benci" and the "Cecilia Gallerani," painted some 30 and 15 years earlier, respectively, the line is seen to pass convincingly close to the leading eye, the right eye in both the earlier

"Art is the Queen of all sciences communicating knowledge to all the generations of the world."

—LEONARDO

Following pages: Florence's celebrated Ponte Vecchio, built in the 13th century, housed butcher shops until the mid-16th century. It was restored under Cosimo de' Medici. The butcher shops were forced out and jewelers allowed to replace them.

MATHEMATICAL UNDERPINNINGS

In the image at right, the golden rectangle (height-to-width ratio 1:1.618) and the associated golden triangle (interior angles 72°-36°-72°) are superimposed on the "Mona Lisa," in order to demonstrate how Leonardo's total preoccupation with mathematics most likely influenced his technique in organizing his composition. First a square is drawn, the height of which is determined by the height of the subject's head; the square is then extended downward by 62 percent to create the golden rectangle. The lower edge of the rectangle is seen to pass through the subject's bodice. Finally, the vertical line is drawn, dividing the frame into two equal halves; this centerline is seen to pass right through the subject's left eye. Herein lies a demonstration of psychologist Christopher Tyler's discovery that in the preponderance of great single-subject portraits through the ages, the painter intuitively positioned one of the eyes of his subject to lie at or near the centerline. One always has to be wary of reading too much into such superimposed geometric figures, but Leonardo, more than any other artist, sought connections between art, mathematics, and science. He had long regarded painting as a science, and his words, "No human investigation can be called true science without passing through mathematical tests," especially resonate in this, his third portrait of a young woman.

portraits. In the case of the "Mona Lisa" it is the left eye. In "La Belle Ferronière," a work whose provenance is somewhat disputable, the centerline passes through the subject's left eye.

When Leonardo painted earlier works, such as the "Adoration of the Magi" and the "Annunciation," the theory of one-point perspective had already been known for 50 years, and Leonardo had mastered it. By the time he painted the "Mona Lisa," Leonardo had learned how to manipulate it in order to produce special effects. For a painting to be considered "lifelike" in the Renaissance, it had to convey a feeling of the subject being alive, rather than rendering an exact

physiognomic or photographic likeness. Since no other images of del Giocondo's wife exist, no one knows what she looked like. But from the canvas she is ready to speak: The outer edges of her eyes have been painted purposefully blurred, creating a sense of ambiguity. Moreover, the landscape forming the backdrop behind her is higher on one side than on the other. Therein lies one of Leonardo's tricks. His design causes the observer's eye to oscillate back and forth across the subject's hypnotic visage.

A more compelling (although slightly more complicated) analysis is offered by Dr. Margaret Livingstone, a professor of neurobiology at Harvard Medical School. In a highly readable book, *Vision and Art: The Biology of Seeing*, Livingstone discusses the essential features of the eye, and the functions of each of these features. In the human eye there are two kinds of photoreceptors—rods and cones—that generate neural signals in response to light for the brain to process. The rods are more sensitive to light and assist us in seeing at night or in dim light. The cones are less sensitive and specialize in seeing in the daytime or in bright light. The retinal distribution of the rods and cones, however, is not uniform, with the cones especially preponderant at the center of the gaze, the rods around the periphery. Accordingly, the acuity and spatial resolutions in the central and peripheral vision are not the same. In order to get a comprehensive picture, we need both a highly focused and fine-tuned image provided by the center gaze, and a more general and coarser image provided by the peripheral vision. A passage from Livingstone's book reads:

> I looked at the "Mona Lisa" as if I had never seen the painting before, and indeed I noticed something I had not seen previously. See if you experience the same thing I did. Look at her mouth, and then look at the background. Look at her mouth again, then her eyes. Look back and forth between her mouth and other parts of the painting. As I did this, I realized that her smile seemed more apparent and cheerful when I was looking away from it, and it seemed less evident when I looked directly at it.... I suggest that her smile is more apparent in the coarse-information component of the image, and is therefore more apparent to peripheral than to central vision. This explains its elusive quality—you literally can't catch her smile by looking at it. Every time you look directly at her mouth, her smile disappears because your central vision does not perceive coarse image components very well.... Mona Lisa smiles until you look at her mouth, and then her smile fades, like a dim star that disappears when you look directly at it

The unevenness of the horizon line in the portrait makes more sense in view of Dr. Livingstone's remarkable analysis. Leonardo forces us to scan back and forth across her face. Again, we ask the question: Did Leonardo do this on purpose, or is it a coincidence? A resounding answer, as you get to know this man's mind, is that there are few accidents or coincidences with Leonardo.

Finally, from the sublime to the preposterous, the "Mona Lisa" has inspired more copies than any other painting in history, but it has also inspired more spoofs. High Renaissance master Raphael copied the portrait, depicting another

"No human investigation can be called true science without passing through mathematical tests."

—LEONARDO

BACKGROUND OF THE "MONA LISA"

"The knowledge of past times and of the places on the earth are both an ornament and nutriment to the human mind."

—LEONARDO

Opposite: The seven-arched Ponte Buriano in Arezzo, Tuscany, appears in the background of at least three Leonardo paintings.

Above: The Balze Rocks in Valdarno, Tuscany, are highly reminiscent of the backdrops of the "Mona Lisa" and the "Virgin and Child with St. Anne."

The rugged countryside Leonardo used as the backdrop in the "Mona Lisa," as well as in the "Virgin and Child with St. Anne" and the "Madonna of the Yarnwinder," is not a single location, but rather a dreamlike imagined vision—an amalgam of the real and the fictitious, just as in the landscape portrayed in that earliest surviving drawing of his beloved Arno Valley (page 18). There are similarities to real sites around the Valley of Chianti, northwest of Arezzo, and to the Balze Rocks, just past the village of San Giustino Valdarno. Geologists explain the origin of the Balze Rocks as formations carved in the Pleistocene period, from roughly 1.8 million years to 10,000 years ago, when repeated periods of glacial activity enveloped much of northern Europe, and at its peak extended south into northern Italy. Among the effects of the Pleistocene were the dramatic climatic changes, the appearance and disappearance of lakes, vast sediment deposits, and the scouring of underlying rocks by material transported by water. The Balze Rocks can be seen in the municipalities of Terranova Bracciolini, Castelfranco di Sotto, Pian di Sotto, Pian di Scò, Laterina, and Regello.

As for the bridge visible just to the right of the Mona Lisa, at the level of her left shoulder, there is a compelling candidate. The locals refer to it as the Via dei Sette Ponti ("road of the seven bridges"), connecting Arezzo and Florence, 50 miles away. The "seven" does not refer to the number of bridges, but rather to a medieval bridge with seven arches crossing the Arno. Ironically, a second and more prominent medieval seven-arch bridge nearby also straddles the Arno. The Ponte Buriano, completed in the mid-1200s, strongly resembles the bridge in the "Mona Lisa" and local folklore confers on it that provenance.

Florentine lady in a pose that is an unmistakable adaptation of the Mona Lisa, but it is surprisingly flat, devoid of the hauntingly hypnotic power of the original. At least two "Nude Mona Lisas" were produced within a decade after the original, including one by Salai. Known as the "Madonna Vanna," it hangs in the Hermitage, St. Petersburg, Russia. In 1919 Dadaist painter Marcel Duchamp painted a thin mustache and goatee on Leonardo's lady; three decades later the defining surrealist, Salvador Dali, could not resist painting his own face on the iconic portrait, replete with his famous long and pointed mustache. By 1977 Colombian painter Fernando Botero fattened up the lady to fit his own distinctive style, then the publishers of *Mad* magazine followed with their own version, presenting Alfred E. Newman sitting in for the Mona Lisa on the cover. Among more recent entries is a large billboard in east Baltimore showing a Mona Lisa in the late stages of pregnancy, with a bold message: "Who's the Father?" (an advertisement for a lab administering DNA paternity testing). Ultimately, the numberless copies and parodies of the world's most recognizable painting represent a decidedly sincere form of flattery.

A sampling of portraits by three great painters demonstrates the centerline principle discovered by Christopher Tyler. Above left: Leonardo, "Lady with an Ermine," late 1480s. Czartoryski Museum, Krakow. Above center: Rembrandt van Rijn, "Self-portrait," 1659. National Gallery of Art, Washington, D.C. Above right: Vincent van Gogh, "Self-portrait," 1889. Musée d'Orsay, Paris

"Virgin and Child with St. Anne"

By the time Leonardo had completed "The Last Supper" at Santa Maria delle Grazie in Milan in 1498, he had begun exploring a design for a painting that would incorporate the Virgin Mary, her mother, St. Anne, the infant Christ, and possibly a young St. John the Baptist. This is precisely the commission Leonardo was awarded by the Servite Friars of the church of the Santissima Annunziata upon his return to Florence in 1500. Upon its unveiling to the public in 1501, the first cartoon for the painting created a sensation in the Florentine art world, after which Leonardo appears to have put the project aside in favor of other interests, commissions, projects, and far-flung employment. This early cartoon does not survive, but a second cartoon for the painting, created probably around 1508, does survive, owned and maintained in conditions of subdued lighting and climatic control in the National Gallery, London. It is a highly prized work in the collection, and still referred to as the "Burlington House Cartoon," after its last private owners.

The "Burlington House Cartoon," measuring 4.6 feet by 3.4 feet, comprises eight separate sheets stitched together. The Virgin is sitting on the knees of St. Anne and holding the infant Christ, who in turn is stretching out to bless the young St. John. In the charcoal and chalk drawing on tinted paper, there is a genuine aura of piety in the Virgin's face, and the notion of an adult sitting on the lap of another adult appears altogether natural in Leonardo's fluid composition. Leonardo mysteriously abandoned this design before preparing it for transfer onto a panel—there are no puncture marks along the outline of the figures, normally a telltale clue that a cartoon has been used in the process of transferring its image to a panel for painting. A contemporary description of the lost 1501 cartoon, which did not portray St. John, is more in keeping with the final painting of circa 1510 than the existing "Burlington House Cartoon." Leonardo bequeathed the cartoon to his loyal apprentice and companion, Francesco Melzi. The cartoon is a work of such sublime beauty and power that art historians have often expressed a preference for it over the finished painting, now housed in the Louvre.

In the painting "Virgin and Child with St. Anne " with St. John missing, the contorted postures of both the Virgin and St. Anne create a sense of tension, making the image of an adult sitting on an adult lap unnatural and heavy. Mary is leaning over to hold the Infant Christ, who is in turn childishly tormenting a lamb. Here the lamb is emblematic of Christ's own later martyrdom. The bemused grandmother, St. Anne, looks on with a smile reminiscent of Mona Lisa. It is a smile that has fueled speculations about whether this is not a reflection of the artist's nostalgia for the woman who passed away while a guest in his home in Milan—probably his mother, Caterina. Such questions will never cease about the works of Leonardo, since he left no notes, but there is no speculation in the fact that the composition of this work inspired other Renaissance masters, including Michelangelo, Raphael, and especially the Venetian painters of the Late Renaissance.

Finally, the background in the painting is evocative of that of the "Mona Lisa," a composite landscape that is half real and half fantasy. But Leonardo quite possibly possessed a genuine photographic memory, and some of this can be simple embellished reality. The bridge in the distance to the left of the trio is most likely Ponte Buriano, a seven-arched bridge in nearby Arezzo that may have served as inspiration for the bridges in the "Mona Lisa" and the "Madonna of the Yarnwinder."

A Contest of Wills

In early 1506, with his commissions still languishing, Leonardo was called to Milan by the French governor, Charles d'Amboise. The Signoria in Florence, concerned especially about lack of progress on the "Battle of Anghiari," acceded to the request of Charles d'Amboise, allowing Leonardo three months leave in Milan. Leonardo's presence was required in Milan in order to satisfy

LEONARDO'S BICYCLE

Early in the 19th century, the first bicycle was introduced. Over the next 80 years, various combinations of a human-powered vehicle were tried out: a unicycle; a two-wheeler, with a large wheel in front, small in the rear; a two-wheeler, with a small wheel in front, large in the rear. By 1895 the optimum design for a bicycle—two equal-size wheels, a chain drive, a pair of pedals, mechanical advantage created by a large gear in front and a small one in the rear—had finally emerged. Four hundred years earlier, circa 1493, an assistant of Leonardo had sketched what must have been the master's idea for the bicycle—two equal-size wheels, a chain drive, pair of pedals, mechanical advantage created by a large gear in front and a small one in the rear. The rough sketch of the bicycle appears in the Codex Madrid I, a collection of drawings that had been lost for centuries—"misfiled" in the Spanish National Library—and then discovered in the 1960s and painstakingly restored.

the judgment handed down in the continuing legal battle over the "Virgin of the Rocks." Presumably Leonardo's partner, Ambrogio de Predis, had replaced the central panel of the altarpiece, which had probably been given to Emperor Maximilian by Ludovico Sforza in 1493. The replacement that the painters agreed to supply the Confraternity was most likely in place before Leonardo left Milan in 1499. Ambrogio, in 1503, in a suit brought on behalf of himself and Leonardo, claimed again that they were owed money by the Confraternity for the finished painting. Finally, in 1506, the decision went against the painters, the court deciding that the painting had not been properly finished, and that a proper finish would require the hand of maestro Leonardo himself, who was then called away from Florence to meet this legal obligation.

He eventually satisfied the Confraternity by touching up their version of the painting, but he also became the guest of Charles d'Amboise and started working on plans for d'Amboise's new villa outside the city. He left drawings for a rustic paradise of sorts. In addition to an airy interior of spacious rooms, in keeping with Leonardo's conceptions of sanitation, he proposes gardens and streams, orchards, bowers, and fishponds— a bucolic dream—topped off by an ingenious water mill that will "create breezes in the summer," provide flow for cooling fountains, and power devices that will generate "continuous music."

Leonardo was obviously having fun and did not want to return, and Charles d'Amboise was reluctant to part with him, even for a short while. There were further exchanges of letters between Charles and the Signoria in Florence, all to no avail. Charles politely but firmly insisted on extending Leonardo's leave of absence; the Florentines, at least to some extent out of deference to the power of the French in Italian affairs, acceded to the extensions. By late 1506 some of the exchanges, particularly on the part of the Florentines, became testy, but Leonardo stayed put into the new year, when King Louis XII, in January 1507, told the Signoria that he had a "necessary need of Master Leonardo da Vinci," and that he was not to leave Milan. Florence could only resign itself to its loss.

Except for a relatively short return to Florence later in 1507 to deal with a legal dispute with his half brothers following the death of his uncle Francesco, Leonardo did not live in Florence again. And by April 1508 he was back in Milan, with all his possessions, including his library, notebooks, and paintings, finished and unfinished—except for the "Battle of Anghiari," abandoned on the wall back in Florence.

Above: Leonardo, "Burlington House Cartoon" for "Virgin and Child with St. Anne," ca 1508. National Gallery, London

Opposite: Leonardo, "Virgin and Child with St. Anne," ca 1510. Musée du Louvre, Paris

LEONARDO'S SKETCHBOOK

FOR LEONARDO, the composition of a painting required understanding of all the elements at a fundamental level: light and shadows on curved surfaces, subtle nuances in hand gestures and facial expressions, the group dynamics involved in tightly knit gatherings—in the quartet of individuals in the "Virgin of the Rocks," the maelstrom of warriors and horses in the "Battle of Anghiari," the lineup of the apostles seated in "The Last Supper"—all original, all influential.

"Study of Drapery Wrapped Around the Legs of the Virgin in the 'Virgin and Child with St. Anne,'" 1508. Musée du Louvre, Paris

"Early Study for the 'Virgin and Child with St. Anne.'" 1508-1510. Galleria dell'Accademia, Venice

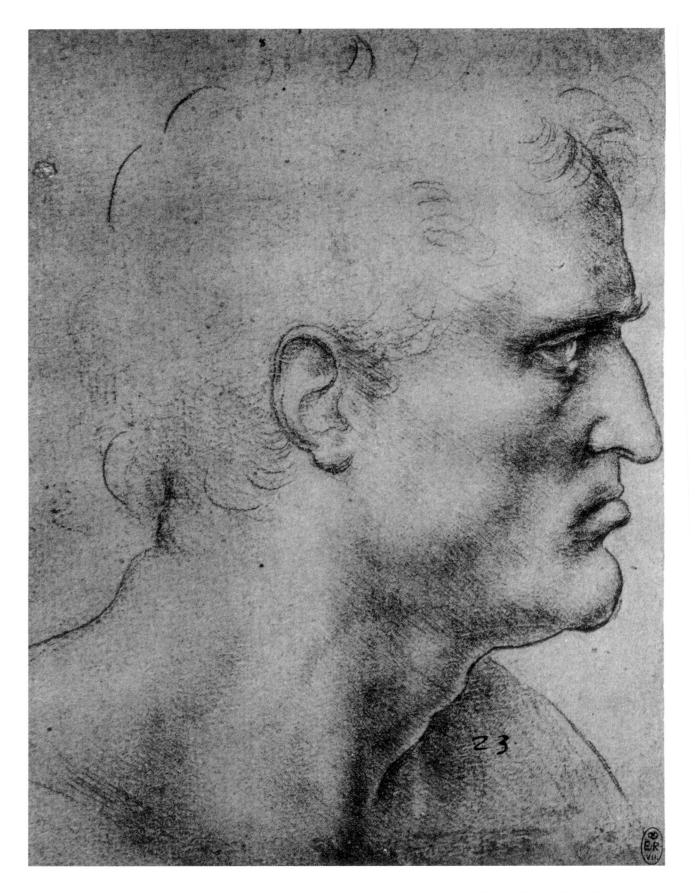

"Study for the Head of Bartholomew, seated far left in 'The Last Supper,'" 1495. Royal Collection, Windsor Castle, U.K.

"Study for Head of a Horse for the 'Battle of Anghiari,' " 1503-04. Royal Collection, Windsor Castle, U.K.

"Study for the Head of St. Anne for the 'Virgin and Child with St. Anne,' " ca 1503. Galleria dell'Accademia, Venice

"Study for the Head of the Infant Christ for the 'Virgin of the Rocks,'" 1483. Musée du Louvre, Paris

Chapter Six

1508-1519
THE FINAL YEARS

ca 1508 Leonardo draws second cartoon (aka "Burlington House Cartoon") for "Virgin and Child with St. Anne."

1508 Leonardo starts to organize notebooks with goal of publication. He leaves for Milan, never to live in Florence again, to taking his paintings, books, and notebooks.

ca 1508-1511 Leonardo studies anatomy with Marcantonio della Torre.

1511 Leonardo leaves Milan for the Melzi family villa in Vaprio d'Adda.

1512 Massimiliano Sforza is restored as Duke of Milan. The Medici return as rulers of Florence. Battle of Ravenna drives the French from Italy.

1513 Leonardo begins "St. John the Baptist." Sponsored by Giuliano de' Medici, he takes up residence in the Belvedere Palace, Rome.

1515 François I succeeds Louis XII as King of France. In Rome, the pope forbids Leonardo further postmortems.

1516 François I invites Leonardo to Amboise, France, as "Peintre du Roi" (King's Painter).

1517 Cardinal d'Aragona's secretary records Leonardo's holdings.

1519 Leonardo dies in Amboise.

While still in Milan during the summer of 1507, Leonardo met a 14-year-old Milanese named Francesco Melzi who became attached to his household as apprentice and personal assistant. Unlike Salai, still part of the entourage, Melzi was well educated for his age and able to be of real assistance to Leonardo in organizing his papers and carrying on his correspondence. He also became a better painter than Salai. Melzi would stay with Leonardo until the end and be named in his will as his literary executor. In that office he inherited the 20,000 pages of notebooks and drawings Leonardo left at his death.

That same summer Charles d'Amboise wrote a letter to the Signoria in Florence granting Leonardo permission to return to the city temporarily to help defend against a legal challenge brought by Leonardo's half brothers against the will of their uncle Francesco, who had died in early 1507 leaving the bulk of his estate to Leonardo. By September 1507, Leonardo was back in Florence for what would be his last extended stay in the city.

Plans for Future Publications

The case was to be decided by November 1, 1507, All Saints Day, but it lingered on. In the first weeks of 1508, Leonardo wrote to Charles d'Amboise to say that the suit would probably be concluded by Easter. When not attending to his lawsuit, Leonardo put these final months in Florence to good use. Most of his possessions, including his now vast collection of notebooks and papers, were still in Florence. He started to organize them with a possible view to eventual publication. Most significantly, he began to put together the manuscript now known as the Codex Leicester, a compilation of his most fundamental scientific investigations and speculations on physics, geology, meteorology, and cosmology. Among specific topics in the collection are the perturbed and unperturbed flow of water and their ramifications for tides and currents. Another is the formation of rocks and mountains. Leonardo of-

Opposite: Three remaining columns of the Temple of Castor and Pollux in the Forum, Rome
Previous pages: View of snow-covered Florence from Piazza Michelangelo

fered prescient speculation on fossilized shells found on mountains having their origins in the seabed, but ending up high on mountains as the result of the land thrusting upward. This is more than 300 years before pioneering British geologist Charles Lyell published his landmark treatise, *Principles of Geology,* where he came to similar conclusions. It is 400 years before German meteorologist Alfred Wegener proposed that the continents gradually drift relative to each other—early murmurings of the theory of plate tectonics, as the explanation for mountain-range building. The actual movement of discrete continental plates on which the continents themselves are embedded was shown conclusively another

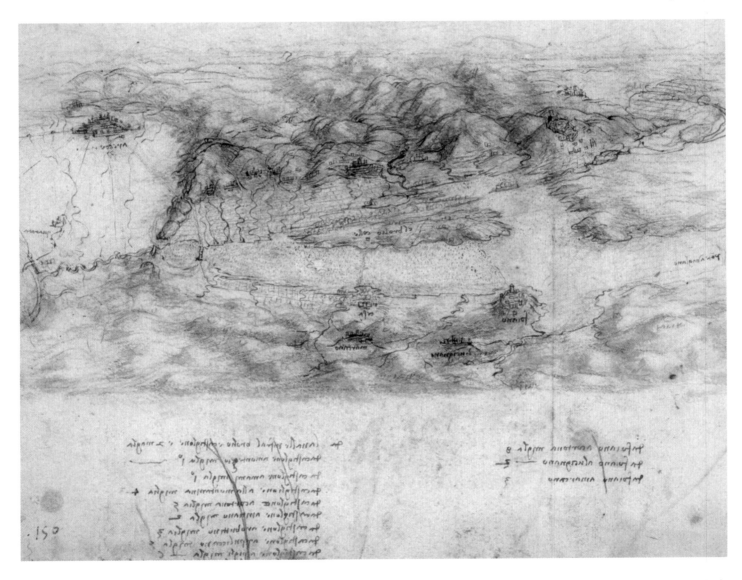

50 years later during the International Geophysical Year of 1957, when British oceanographers determined that the ocean floor of the Atlantic was indeed spreading, and even measured the speed of the separation of the continents.

Again in geology, while Leonardo incorrectly accepted the prevailing belief (handed down by natural philosophers of antiquity) of the existence of water deep below the Earth's surface, and also incorrectly attributed the variegated tones on the moon to the existence of seas on the moon's surface—just as Galileo would do over a century later—Leonardo correctly explained that it is "earth shine" (sunlight reflected by the Earth) that reveals the full circular shape of the moon when direct sunlight hitting the moon from an oblique angle illuminates it only partially (for example creating only a crescent moon). In this explanation he prefigures Kepler by over a century. As he gathered this material, Leonardo made notes to himself about putting it all in order, indicating this was only a first effort to bring together the subjects of interest.

Anatomical Studies

During the early months of 1507, Leonardo also began a serious study of human anatomy. These investigations went beyond the preliminary and cursory. Though not published in his lifetime, his anatomical studies were the most thoroughgoing of all his scientific researches, consuming his attention and curiosity over the following five or six years.

Verrocchio's admonition to his apprentices—build the body from the inside out—resonated with Leonardo. Verrocchio could have had no idea, however, of the lengths Leonardo would go to understand human anatomy. Leonardo scholar Sherwin Nuland, professor emeritus of surgery at Yale University Medical School, and one of the great anatomists of our time, confesses to "an idolatrous admiration" of Leonardo and calls him the "finest anatomist ever."

During his earlier decades in Milan—from 1482 to 1499—Leonardo had done dissections, but mostly on animal carcasses rather than human cadavers. His early views on the human body were shaped by the writings of second-century Greek physician Galen or based on medieval anatomical beliefs. At that time, Leonardo accepted Galen's view that blood originated from the liver. He had not made any pronouncements on the heart as a muscle, and his drawings of the heart of a calf and the embryo of a cow, though rendered in exquisite detail, were not significant for the understanding of human anatomy. Although his earlier skull drawings are remarkable, his thoughts on the site of thought, or mental processing, rarely went beyond speculation or introspection. Little evidence exists that he performed postmortems on complete human cadavers in this earlier period.

During Leonardo's final months in Florence, roughly late 1507 to early 1508, his mastery of human anatomy began to evolve. He started a program of careful dissections of human cadavers, recorded in finely articulated drawings and accompanying notes that revealed many structures and details never before described. Two human dissections performed at the hospital of Santa Maria Nuova were particularly significant. The first was carried out on a recently deceased old man who did not know exactly how old he was but knew that his "grandson was fifty-eight"; Leonardo guessed this would make the man more than a hundred years old. "[T]hus while sitting on a bed in the hospital of Santa Maria Nuova in Florence,

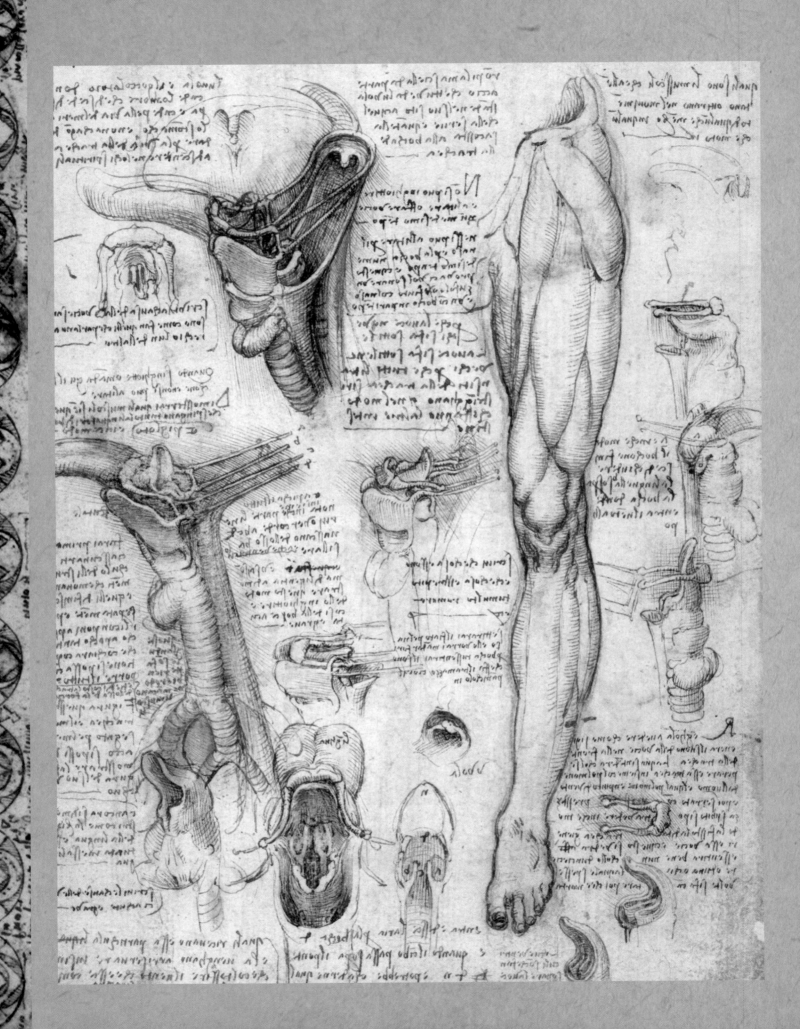

MEDICINE AND ANATOMICAL STUDIES

"Human subtlety ... will never devise an invention more beautiful, more simple, or more direct than does nature, because in her inventions, nothing is lacking, and nothing is superfluous."

—LEONARDO

Above: A 1714 copper engraving, artist unknown, of a Renaissance anatomical theater

Opposite: Leonardo, "Study of the Anatomy of the Throat and Leg," ca 1510-11. Royal Collection, Windsor Castle, U.K.

In what we regard as our enlightened age, we take for granted the existence of institutions that care for the physically and mentally infirm, where physicians and nurses practice basic science, where myriad tests and surgeries are performed, where complex medical imaging devices peer into our bodies. In medieval times, those hapless individuals who suffered from serious infirmities were often the discards of society, doomed to live on the streets in squalid conditions, and survive by begging. The institution known as the *ospedale* (literally "hospital") appeared in Italy, initially as a hospice, not for providing cures for the ill, but for providing shelter for the infirm. Conditions of hygiene were deplorable—two or more patients could occupy a single bed, changing bed sheets was never considered, and, in women's wards especially, chickens might be kept. One medieval woodcut shows a pair of patients in one bed, and in a corner of the room, a carpenter creating wooden coffins. Far more important

than any medical care was the presence in the room of the Holy Spirit—a painting showing Christ or the Virgin, a view of a chapel (and in Rome, perhaps even a distant view of the Vatican).

Medicine as a formal subject of study was introduced in universities in Italy in the 13th century—first at Bologna, then subsequently at Padua, Naples, Siena, Rome, Pavia, and Turin—then spread to the rest of Europe: Paris, Oxford, Cambridge, Salamanca, etc. Anatomy and cures for illnesses were taught strictly according to Galenic teachings, handed down from the second century A.D. Medieval and Early Renaissance woodcuts depicting anatomical dissections always show the collaboration of two individuals—the professor of surgery, wearing a long red robe, emblematic of his high social and professional stature, and the surgeon, shown wearing a short tunic, symbolic of his lower stature. The surgeon, who also doubled as barber, would carry out the postmortem quite literally in an operating pit, while the professor of surgery, sitting in a high chair with a Galenic textbook in his hand, supervised and explained to the students what they were seeing. In the most famous medical school of all, Padua, the students observed the postmortem from the concentric oval balconies of the

operating theater, which they lined shoulder to shoulder.

Medical historian Rina Piana reveals the intellectual dishonesty that often prevailed in these sessions, where observation was subordinated to dogma. For example, the surgeon would display an arrow-straight humerous (the bone of the upper arm), while the professor of surgery would explain an optical illusion at play, that "it is one of nature's jokes to make the bone seem straight when in reality it is curved [just as Galen had described it]." (Never mind that Galen had based his description on the humerous of a pig.) The professor would reiterate that Galen was the true authority, and more so than nature. This is the sort of nonsense that Leonardo would have rejected out of hand had he been "subjected" to a formal education. The most vociferous proponent of learning from nature, Leonardo had to see and touch for himself. Piana points out that at this time (honest) anatomical studies were being carried out not in medical schools, but by the artists, some of them, most prominently Leonardo and Michelangelo, paying church authorities in order to have access to corpses. In the wider context of science and not just anatomy, it was the artist who taught the scientist "how to observe."

without any movement or sign of distress," Leonardo wrote, "he passed away from this life. And I made an anatomy of him in order to see the cause of so sweet a death. This anatomy I described very diligently and with great ease because of the absence of fat and humors which much impede knowledge of the parts."

The man's lean body allowed Leonardo to compose his most complete record of a single postmortem; however, some of the resulting drawings were actually composed not at the hospital, but later, based on his notes, after he returned to his studio. Thus, in many of these drawings, he was depicting what he remembered seeing rather than what he was actually observing while drawing. In fact, the drawings of this postmortem are secondary to Leonardo's notes, which contain the first known descriptions of the effects on the organs of "cirrhosis of the liver, arteriosclerosis, calcification of vessels, coronary vascular occlusion, and capillary vessels." Leonardo concluded in the case of the old man that the cause of death was "a weakness through failure of blood and of the artery that feeds the heart and the lower members which I found to be very parched and shrunk and withered." He appears to understand the circulation of the blood more than a hundred years before the discoveries of William Harvey (1578-1657).

Around the same time, in a subsequent autopsy on a child, Leonardo found the same vessels to be clear of obstruction and the vessel walls supple. What Leonardo had described in the coronary arteries of the old man was atherosclerosis—hardening resulting from plaque formation on a blood vessel's interior walls—and arterial obstruction, a diagnosis of heart disease made hundreds of years before either of the symptoms was recognized by physicians. Medical historian Kenneth Keele, who made some of the most incisive studies of Leonardo's anatomical research, pointed out that the understanding of atherosclerosis represented the fruits of a multilateral approach that, more than any other, defined his scientific inquiry. In this instance, Leonardo systematically studied and connected three areas: hydrodynamics, the anatomical changes wrought by aging, and the effects of nutrition, and at a more fundamental level, the physics underlying biology. His lifelong preoccupations with the flow of water had led him early on to invent the use of markers—weighted floats, bits of paper and cork, fine grass seed, and dyes—that could be observed while the water in which they were suspended was flowing, or when it was pumped through clear glass tubes. In our modern age, when medical imaging technology can peer into

Opposite: Leonardo, "Study of Details of Vital Organs: Cardiovascular System, Spleen, Liver…," 1510. Royal Collection, Windsor Castle, U.K.

Below: Leonardo, "Study of the Anatomy of the Shoulder," 1510. Royal Collection, Windsor Castle, U.K.

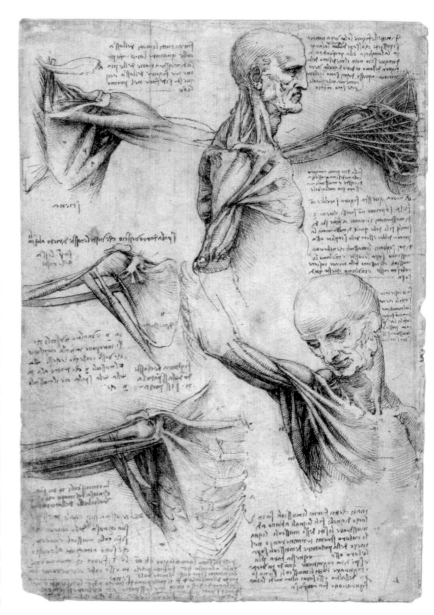

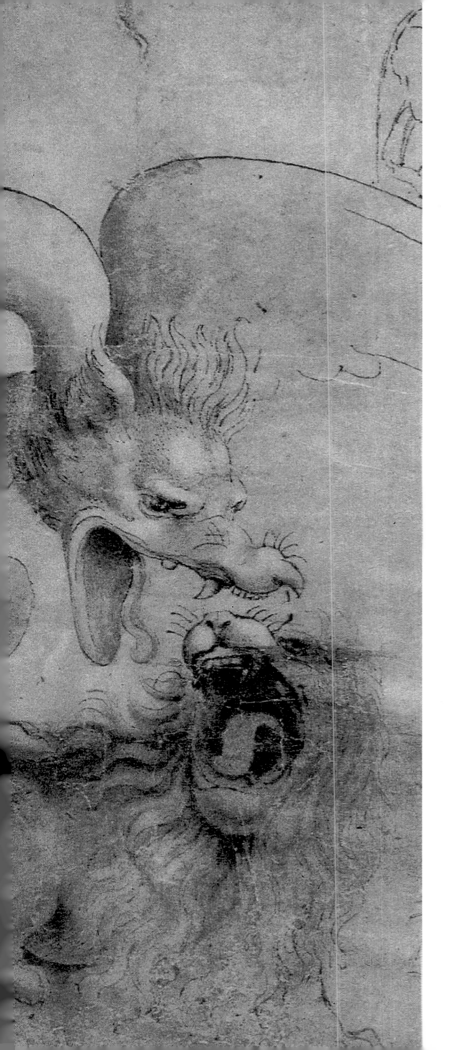

"I wish to work a miracle."
—LEONARDO

Leonardo, "Battle Between a Dragon and a
Lion," ca 1481. Gabinetto dei Disegni e delle
Stampe degli Uffizi, Florence

LESSONS FOR CARDIAC SURGERY

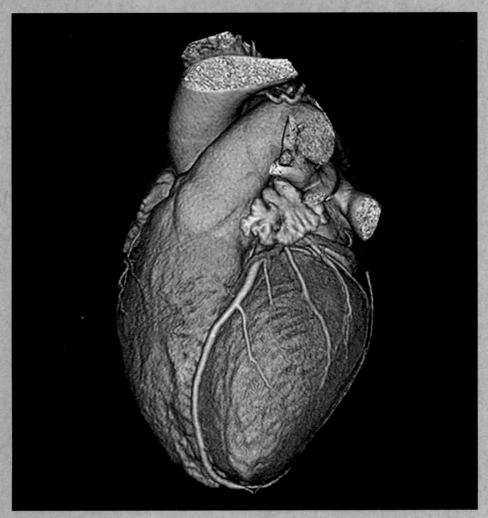

"How could you describe this heart in words without filling a whole book?"
— LEONARDO

reasons, the leaflets of the diseased valve lack the normal symmetry in shape and in motion. The leaflets do not meet properly and the valve leaks. As this happens, the LV dilates to compensate for the fact that some of the ejected blood flows in the wrong direction. Under severe conditions, blood regurgitates through the LA back into the lungs, yielding a state of congestive heart failure. Typically, symptoms include increasing shortness of breath and fatigue. Until Wells's new surgical approach, surgeons had repaired a floppy valve by cutting out the prolapsing segments and narrowing its orifice. This, however, inevitably restricted the orifice and blood flow, especially during strenuous exercise.

Leonardo's approach to solving problems became a lesson for Wells, who reassessed the cause of the problem. Leonardo had discussed the concept of "minimal energy surfaces," or boundary surfaces between two pressurized chambers organizing themselves in a manner that allowed the uniform dissipation of the forces exerted on them. Wells arrived at an alternate solution, embracing a more physiologically (naturally) functioning valve. After repair, the normal size of the valve in its open position would not be reduced, its structure and function would remain unchanged, and the patient could return to vigorous exercise more quickly. By early 2008, Wells had successfully treated 300 patients.

Leonardo has also inspired surgical technology. Intuitive Corporation of California invented a revolutionary machine that performs a variety of minimally invasive surgical procedures, one that they market as the "Da Vinci Robot."

Above: A CT (computed tomography) scan image of a normal human heart

Opposite: Leonardo, "Studies of the Heart," ca 1510-13. Royal Collection, Windsor Castle, U.K.

In late 2005 a number of stories appeared in the media about a leading British surgeon who had taken Leonardo's lead in revising a procedure commonly used in modern heart surgery. Dr. Francis Wells, a widely published cardiac surgeon in Cambridge, England, introduced a novel approach for repairing the heart's mitral valve. Leonardo's general intellectual methodology, gleaned from the study of his notes, drawings, and observations on the heart, inspired Wells to revisit the conventional surgical approach to mitral valve repair.

The mitral valve (MV), on the left side of the heart, is one of the four valves of the heart. Separating the left atrium (LA; left upper chamber) from the more powerful left ventricle (LV; left lower chamber), it permits unidirectional blood flow from the LA to the LV. The MV consists of two fibrous leaflets, or flaps, that are attached at the fibrous annulus, or ring, of the atrioventricular junction. During the quiet phase of the heart cycle—also called diastole—blood flows passively from the LA to the LV through the MV. When the LV begins to contract, pressure in the chamber builds and the MV passively closes. Blood is then forced across the aortic valve and into the aorta, which in turn distributes it to all of the body's organs. The developed ventricular pressure might cause retrograde flow of blood across the MV if not for two muscular projections called the papillary muscles that reside within the LV cavity. These muscles are attached to the tips of the MV leaflets and, by contracting simultaneously with the LV, they prevent the valve from failing. It is an ongoing tug-of-war that a normal valve never loses.

In a condition called mitral valve regurgitation, however, it does lose. For a variety of

LOGARITHMIC SPIRAL

In his studies on turbulent flow in water, Leonardo discovered the shape of the whorl, a special spiral known to modern mathematicians as the logarithmic spiral. In his drawings, he employed the spiral in depicting the embryo of a cow and the curls in the hair—most prominently on the head of the "Portrait of a Musician" and the angel in the "Virgin of the Rocks." As an urban designer he drew circular staircases, ostensibly for their graceful beauty, but more to eliminate corners where the public could urinate. Leonardo may have inspired the famous double-spiral staircase—a double helix—in the Château de Chambord of François I. Construction on the château in the Loire Valley was begun in 1519, and its staircase would be emulated widely. The staircase Giuseppe Momo designed for the Vatican Museums in 1932, one of the most recognized in the world, was certainly one of those inspired by Chambord. In interweaving a pair of circular ramps—one spiral ramp for ascending, the other for descending—he solved the logistic problem of ushering in and ushering out thousands of museum visitors daily. Viewing a circular staircase from the top, as is the case in the photograph on pages 250-51, creates the image of a mathematical logarithmic spiral. This is the spiral that nature prefers above all others. It is seen in the chambered nautilus, the horns of a ram, the winds in a hurricane, the water in a whirlpool, and the stars in a spiral galaxy.

human and animal bodies, employing dyes as markers is commonplace. But five hundred years ago, Leonardo was arriving at his conclusions indirectly, by studying analogous systems. He had long felt that the human body was a microcosm analogous to the macrocosm representing the Earth. That is the sense in which he wrote: "We may say that the Earth has a vital force of growth, and that its flesh is the soil; its bones are the successive strata of the rocks which form the mountains; its cartilage is the porous rock, its blood the veins of the waters. The lake of blood that lies around the heart is the ocean. Its breathing is the increase and decrease of the blood in the pulses, just as in the Earth it is the ebb and flow of the sea."

With old rivers, like the Arno, Leonardo found that the meandering path leads to increased meandering, with erosion on one bank causing silting on the other, and he saw this as a physical model for the human body. In the elderly, the vessels of the cardiovascular system displayed stretching, constrictions, and dilation unlike the situation in the young. He wrote: "I have also found in a decrepit man the mesenteric vessels constricting the passage of blood, and doubled in length." Leonardo recognized that flow velocity varied inversely with the dimensions of the river, that the angle of incidence in a fluid striking a canal wall equaled the angle of reflection, that ordered systems, in time, increased in disorder—all established

fundamental laws of physics. Keele also reported Leonardo's words: "Vessels which by the thickening of their coats in the old restrict the transit of the blood, and from this lack of nourishment the old, little by little with a slow death destroy their life without any fever; and this happens through lack of exercise since the blood is not warmed." As Keele pointed out, "Leonardo embarked on [the] dissection of the old man with the announced intention of finding out 'the cause of so sweet a death.' It was therefore primarily a postmortem in search of pathology. The fact that it eventually constituted an incomparably rich research into normal anatomy was secondary."

The dissections he performed on these two freshly deceased cadavers—the old man and the child—stood in dramatic contrast to others he performed, mostly in fetid chambers, on cadavers in various states of decay and also described in great detail. The descriptive narrative, however, in those earlier cases was less significant than the detailed anatomical drawings. In these drawings Leonardo was struggling for a technique that would allow a thorough display of the body's structures, both their surface detail and their inner workings. He settled on showing features from as many as eight different angles, in cross section and in multiple cutaway views that successively peel back layers of muscle and tissue to reveal the levels of structure. Anatomical drawings of the time were crude compared to Leonardo's drawings; depictions of the female body in the literature of that time could have been mistaken for schematic drawings of the anatomy of a frog.

In his anatomical studies, it was the eye that fascinated Leonardo more than any other part of the body. The "window to the soul" had to be understood at all levels: its structure, its connections to other organs, and its precise operation as light

Above: Leonardo, "Studies of Flowing Water," ca 1508-1510. Royal Collection, Windsor Castle, U.K.

Opposite: Leonardo, "Study for Trivulzio Monument," ca 1511. Gabinetto dei Disegni e delle Stampe degli Uffizi, Florence

Following pages: This double-helix staircase in the Vatican Museums, designed by Giuseppe Momo in 1932, may have been indirectly inspired by Leonardo's designs for spiral staircases.

Omme ou pre
mier liure ay
parle du come
cement de cete
et durame des
anciens fais
des gregois
pour la mati
ere que iar

emprinse ordinairemet conduire. me
conuient en cest second liure de la cr
acion de troye faire mencion. et de la
mortelle haine qui entre les troyens
fu. et entre les gregois. Car ainsi coe
entendre pourriez si cruellement le gre
demenerent que finablement en fut
troye destruicte ⁊ du monde y mourir
grant partie. Si vouldray ie mostrer

la pugnacion que dieu pour celui fait
aux gregois de puis enuoia. Et aussi
come des troyens furent depuis pluhe
urgnes peuplez ⁊ habitez. et comme
gens sans vraye congnoissance au
de dieu pluseurs terres conquirent
Car pour celui temps ne estoit la
sainte foy encores en lumiere. ne de
long temps apres Iusqes au benoist
aduenement de Ihucrist pour la quele
chose tout le monde estoit en tenebres
et aduegle des obscuritez denfer
Car tout ainsi que dieu par son doly
voulour nous a ordonne la clarte du
ciel pour nous enluminer veult il q
la lune et chune des autres planetes
selon leur naturel cours oeuvre en so
espoure est il q toutesfois q cronctio

THE STATUE OF LAOCOÖN

The story of the Trojan War and the final victory of the Greeks by means of the ruse of the hollow wooden horse is told by the mythical Greek poet Homer in two great epic poems, *The Iliad* and *The Odyssey*. The Greeks had sailed to Troy in their thousand ships and laid siege to the city in order to recover the beautiful Helen, wife of Menelaus, King of Sparta. She had been kidnapped by Paris, a prince of Troy, who claimed her as a prize promised to him by Aphrodite, after Paris had awarded Aphrodite the Golden Apple intended for the fairest of the goddesses.

After ten years of fighting and the deaths of the greatest heroes on both sides, an intractable impasse developed. So the Greeks tried a new idea suggested by wily Odysseus. They constructed a colossal wooden horse, and with a contingent of Greek warriors, including Odysseus, concealed inside, they rolled it in front of the gates of Troy. The rest of the Greeks boarded their ships and sailed away, but they sailed just around the nearest headland, where they remained, waiting for the signal to return once the concealed warriors were inside the city and able to open the gates to the whole Greek war host. Waking to the puzzling sight of the huge horse outside their city, the Trojans immediately began to argue. Some interpreted the horse as a token of Greek respect, extended after resigning themselves to the futility of breaking the will of the Trojans. A few disagreed, Laocoön, a priest of Poseidon, the most vehement among them. He thought the horse was some sort of trick or fraud. Laocoön's words might have convinced the Trojans to reject the horse, to set it on fire outside the city gates. But the goddess Athena, who along with Hera hated the Trojans after the slight of not receiving the Golden Apple, released a pair of sea serpents to attack the irritating priest and his two sons, Antiphantes and Thymbraeus. The Trojan leaders interpreted this as an unmistakable sign

Opposite: King Priam greets his son Paris and Helen in a scene from the circa 1420 illustrated manuscript Chronique de la Bouquechardière, *by Jean de Courcy (1350-1431). Russian National Library, St. Petersburg*

Above: Hagesander, Athenodoros, and Polydorus, "Laocoön," ca 42-20 B.C. Vatican Museums

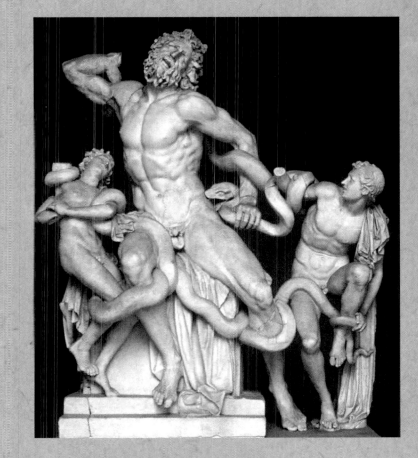

from a benevolent god showing the priest's mistake. They dragged the horse inside the city walls, and, in the process, made one of the most famous blunders in military history.

According to Pliny the Elder (ca A.D. 23-79), in his *Natural History*, the statue, depicting the attack of the serpents on Laocoön and his two sons, is the masterpiece of three sculptors from the island of Rhodes—Hagesander, Athenodoros, and Polydorus. It was probably carved between 42 and 20 B.C. Pliny placed it in the palace of Emperor Titus and called it the greatest work of painting or sculpture the world had yet seen.

The statue disappeared for more than a thousand years before it was rediscovered on January 14, 1506, during excavations of a vineyard near the basilica of Santa Maria Maggiore in Rome. Pope Julius I purchased it and put it on display in the Belvedere Palace, where Leonardo lived between 1513 and 1516. Interestingly, although Leonardo no doubt saw the sculpture often, and must have been as impressed by it as was his great rival Michelangelo, he made no surviving comments about the masterpiece.

Parts of the right arms of Laocoön and both his sons were missing when the sculpture was recovered. The pope held a contest among sculptors to propose how the statue could best be restored. Most proposals were in favor of extending the arms of the priest and his sons outward and upward—a sort of heroic gesture of defiance. Michelangelo dissented, insisting that originally the arms must have been bent back behind the shoulders, emphasizing the tight restraint in which the serpents held the three human figures and preserving the pyramidal structure of the composition. But for more than 400 years the statue was displayed with the Renaissance addition of the extended arms. Between 1799 and 1816 it could be seen in the Palais du Louvre in Paris, where it had been transported by Napoleon after his successful campaigns in Italy. In 1816 the British mandated that the statue be returned to Italy. When the original arm of Laocoön was recovered in the 20th century, Michelangelo was proven correct. The arm of the priest was bent back behind the shoulder just as he had insisted.

Below: An illustration from Johannes de Ketham's Fascicculus Medicinae, *published ca 1490*

Opposite: Leonardo, "The Great Lady," the principal organs and arterial system of a woman, ca 1510. Royal Collection, Windsor Castle, U.K.

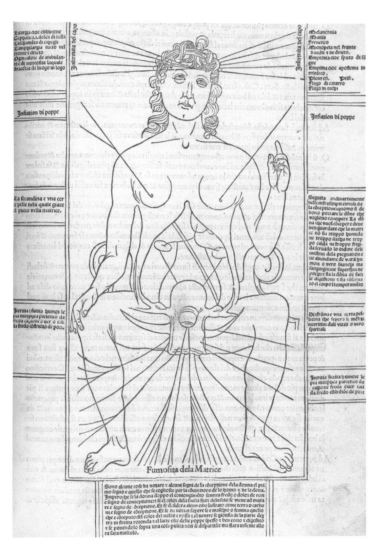

traveled through it. In order to understand the anatomy of this aqueous, mysterious organ—especially difficult to take apart because of its fluid interior—Leonardo had to perform minute dissections. To do so, he invented the technique of holding the eyeball fixed in a glutamate formed by a hard-boiled egg. (The eyeball was immersed in egg white and hard-boiled within it.) The technique of embedding the eyeball in a coagulant before slicing is routinely used today. Leonardo may have performed some of his first work on the human eye while still in Florence during 1508, for in that year he records some of his findings. Debunking the prevailing view of the eye as an active organ that sends out invisible rays, he proposed that the eye is a purely receptive organ that "sees" by means of reflected light.

Back in Milan

Sometime around Easter 1508, Leonardo returned to Milan with all his belongings. Except for two brief visits—en route to Rome in 1513 and during an official visit in 1515, accompanying a papal entourage—he never entered Florence again. In Milan his work on anatomy continued. Simultaneously, he must have worked on completing the "Virgin and Child with St. Anne" and the lost "Leda." He may even have continued to modify the "Mona Lisa." He also participated in staging more court pageants and extravaganzas, most notably a victory pageant for his patron King Louis XII when the king returned to Milan after beating the Venetians in 1509. A mock battle was staged between a lion, representing the French, and a dragon, representing the Venetians. The accounts of contemporary witnesses, as usual, speak of the event with awe. Leonardo's studio continued to produce generic works, such as the "Madonnas," but these were almost exclusively by the hands of the apprentices.

Leonardo also did some preliminary work in hopes of winning the commission to do the funerary monument for condottiere Giangiacomo Trivulzio. In 1504 Trivulzio had set aside 4,000 ducats for the design and construction of his monument. Leonardo's rough sketches for the proposed monument in 1511 include an arch and plinth topped by a bronze horse and rider. He also estimated the cost of the materials and the casting of the horse. But nothing more came of it, and there is no evidence that documents were signed or agreements entered into; further, Trivulzio did not die until 1518, just months before Leonardo's death.

Pageants and funerary commissions were a sideshow for Leonardo. He enjoyed a stipend from the French king that required nothing particular in return. Indeed, he was free to pursue whatever interests struck his fancy—which, at this time, was science. He also continued the attempt to reorganize and bring order to his notebooks with an eye to eventual publication. In late 1508 he

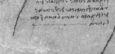

began a notebook collecting his studies and thoughts on water—waves, currents, vortices. In the same notebook he speculated about cosmology and the origins of Earth, disagreeing with classical authorities about the nature and size of the sun. And his interest in flight—never really abandoned—resurfaced in the form of fascination not only with birds but also with bats. He had decided that the wing of the bat was the best model for a flying machine, claiming that it was more stable than the feathered wing of a bird because it allowed the bat to fly upside down and change direction at sharp angles. His notebooks also include geometry problems, including work on the calculation of cube roots. And the study of anatomy

continued. He compiled a notebook dealing exclusively with the eye and vision, writing that "in the winter of 1510 I expect to complete all this anatomy." In this case he was unmistakably looking toward publication, for he left instructions that the drawings were to be reproduced by copper engravings rather than by the crude method of woodcuts.

Learning from the Master

None of his reordering and compiling was ever finished to his satisfaction, however, and nothing was published. And he continued to be dissatisfied with the completeness of his anatomical investigations. His study took a new direction as he became acquainted with a talented young professor of anatomy, Marcantonio della Torre, who had recently moved from the University of Padua to the University of Pavia. The collaboration with della Torre, who was half his age, galvanized Leonardo's dedication to the pursuit of this field of study. There is some disagreement among Leonardo scholars as to whether della Torre was the teacher, directing Leonardo's dissection procedures, or an equal partner deriving as much from Leonardo's work as he contributed. A fair assessment 500 years after the fact might be that della Torre indeed gave structure to Leonardo's methods, but Leonardo—with his innovative mind, his artist's dexterity, and, above all, his independence of thought—took it from there. Just as he had always been a scientist doing art, he was also an artist doing science. The collaboration with della Torre, however, lasted no more than three or four years; in 1511 the young professor died, a victim of the plague.

Leonardo's anatomical studies dating from 1510 to 1513 revealed a new modus operandi that prefigured modern scientific methodology. In earlier studies his method had been to interpret what he saw in the light of prevailing theory; those studies represent a synthesis of his observations. His new method reflected a more open mind, a greater intellectual honesty, and a freedom from prevailing notions. He first recorded what he saw, then sought out functions and causes. Leonardo recognized the heart as a muscle, he described the atrial appendages (auricles), the movements of diastole (the rhythmic expansion of the chambers as they take in blood) and systole (the contractions as the blood is pumped into the arteries), and he understood clearly the functioning of the valves—allowing blood flow in only one direction. The number of human dissections he carried out grew in number from ten in 1508 to "more than thirty" by 1515, according to Leonardo's own claim to a visitor during his retirement in France.

Rome

In 1511 Leonardo once more found himself at the mercy of Italian politics. And once more he survived. On March 10, 1511, Milan's French governor, Charles d'Amboise, died. Leonardo continued to receive his stipend from the French king, but a new menace loomed. Pope Julius II, who had allied himself with the French against Venice, now favored ejecting foreign forces from Italy. By

Above: Leonardo, "Bacchus," ca 1513. Musée du Louvre, Paris. The subject in this painting was initially identified as St. John the Baptist.

Opposite: Pietro da Cortona, "Michelangelo with Pope Leo X," ca 1620. Casa Buonarroti, Florence

late 1511, the Sforza family and their Swiss allies were once more threatening to oust the French from Milan, and on December 29, 1512, Massimiliano Sforza, legitimate son of Ludovico Sforza, and his half brother Cesare, son of Cecilia Gallerani, marched back into Milan. Leonardo could not have welcomed these developments. In the eyes of the Sforza, he had been collaborating with the enemy for most of the past 13 years. Leonardo headed for the countryside just outside Milan and the seclusion of the villa and estate of Girolamo Melzi, father of Francesco Melzi.

Back in Florence, politics was in a similar state of upheaval. In the summer of 1512, the Medici returned to the helm of Florence's ship of state after an absence of 18 years. Giuliano and Giovanni de' Medici, sons of Lorenzo de' Medici, were back in the city, and chief magistrate Soderini was forced to flee. Giovanni, a powerful cardinal, became Pope Leo X in March 1513, after the death of Julius II. Giuliano was initially in charge in Florence, but after his brother was elected pope, he was called to Rome and put in charge of the papal army, and a trusted nephew was left to deal with Florence. Sometime shortly after his arrival in Rome in the summer of 1513, Giovanni, influenced by Giuliano, an aficionado of Leonardo's work, issued an invitation to Leonardo to join him in Rome as a member of the Medici court, and on September 24, 1513, Leonardo and his entourage, including Francesco Melzi and Salai, were on the road to Rome.

Leonardo was given apartments in the Belvedere Palace in the precincts of the Vatican. It occupied a place in the midst of an enormous and largely wild garden, which must have suited Leonardo. The villa had been built 30 years earlier as a papal residence by Pope Innocent VIII. Alterations were made to the rooms to accommodate the needs of Leonardo and his entourage. Everything no doubt was cozy and comfortable—and, as with his French patrons, little was required of Leonardo other than to be present as an adornment to the court. Leonardo continued his interminable efforts to organize his notebooks, and anatomical studies occupied much of his attention.

Little is known about Leonardo's social life or his contacts with other figures of the art world during this period in Rome. His old friend Bramante, architect of the new St. Peter's Basilica, was now the leading practitioner in town, heading up the papacy's efforts to restore the Eternal City to its former glory, an effort that was uncovering many long-buried artifacts of the Roman past as foundations were being laid for new buildings. Most prominent among these treasures was a magnificent first-century B.C. Greek sculpture of the mythical Trojan soothsayer Laocoön in the clutches of a pair of sea serpents, unearthed near the basilica of Santa Maria Maggiore. Michelangelo confessed to a great admiration for the piece. Leonardo makes no mention of it, or of any of the other Roman art and architecture then coming to the surface, although it is unlikely that he would not have been keenly interested in these discoveries. Nor does he mention any of the major figures of the Italian art world then in Rome. Michelangelo was there, as was Raphael, whom Leonardo had met earlier in Florence. Baldassare Castiglione, author of *The Book of the Courtier* and close friend of Leonardo's patron Giuliano de' Medici, was in Rome, and like Leonardo, close to the life of the court. But none of these figures are mentioned in his notebooks of this period. Leonardo appears to have stuck close to his desk while making time for one last painting, possibly a commission from Pope Leo X.

PUSHING SEXUAL BOUNDARIES

As in Leonardo's earlier works, "St. John the Baptist" is entirely open-ended, but unlike the others, it is suffused with disconcerting ambiguities. Instead of the gaunt, stoic, and reclusive character one usually associates with the biblical St. John the Baptist who supposedly survived on honey and locusts in the wild, this is a character with soft effeminate features, entirely androgynous. In fact, it is unmistakably the same hermaphrodite that Leonardo had sketched in the lewd "Angel in the Flesh." A painting from the period, known as the "Nude Mona Lisa," clearly inspired by Leonardo's androgynous sketches, hangs in the Hermitage in St. Petersburg, Russia.

So why does Leonardo depict St. John the Baptist with such androgynous features? And is this not the same ambiguity that he had created by rendering the Apostle John—sitting on the right-hand side of Christ in his masterpiece gracing the wall of the Santa Maria delle Grazie—with distinctly feminine features and demeanor, so that 500 years after Leonardo painted it, Dan Brown, in his *Da Vinci Code*, would offer "him/her" up as Mary Magdalene, the wife of Jesus Christ?

> "I made an anatomy of him
> in order to see the cause of
> so sweet a death."
>
> —LEONARDO

"St. John the Baptist"

"St. John the Baptist," the last known painting completely by Leonardo's hand, exudes clear Leonardesque qualities: First, it is entirely experimental in its approach, presenting the subject seeming to emerge from total darkness with features so fully rounded that no bas-relief, indeed no statue, could convey depth more effectively. The subject's hair features the artist's signature curls, evocative especially of the hair of the "Musician." Also, as in two drawings—"Angel in the Flesh" and a study for "St. John the Baptist"—the subject is pointing an index finger toward heaven. In the "Burlington House Cartoon," the cartoon once intended for the "Virgin and Child with St. Anne," the figure of St. Anne is seen with the index finger of her left hand pointing up. Earlier, when Leonardo painted the first version of the "Virgin of the Rocks," the angel was depicted pointing at the infant figure of John the Baptist. The message conveyed by these gestures is that of eternal salvation through St. John, a popular theme in religious paintings of the time. Finally, the face of the subject bears the enigmatic smile that defines the "Mona Lisa" and is prominent in St. Anne in the "Burlington House Cartoon." In his partially obscured left hand, St. John the Baptist holds a cross made of reed and wears an animal hide. This garment, however, reveals the hand of an anonymous artist and was probably added to the painting at a later time.

Leonardo had pursued his anatomical studies and continued his dissections of human cadavers in Milan and then in Rome until he clashed with Giovanni degli Specchi, a mirror-maker in Rome. The relationship between the two is not entirely clear, but it is probable that Leonardo—at the time conducting experiments using concave mirror constructions to concentrate sunlight in order to produce intense heat—was doing business with Giovanni. Leonardo accused him of spying and professional jealousy and of interfering with his apprentices. Giovanni responded by denouncing Leonardo for performing unauthorized dissections on human cadavers. The renunciation led to investigations by a commission organized by Pope Leo X; in 1515 the commission issued a decree forbidding Leonardo to undertake further anatomical work.

Final Journeys

In October 1515, Leonardo took part in a papal entourage that traveled from Rome on a state visit to Florence—Leonardo's last visit there—and to Bologna. The pope made his grand entrance into Florence on November 30 and stayed for a week. Part of the purpose of the visit of the Medici pope to his native city was to convene a meeting of artists and architects to begin plans for a new and more magnificent Florence. Some of the meetings must have taken place in the Council Hall of the Palazzo della Signoria in front of the unfinished "Battle of Anghiari." We do not know how Leonardo reacted to this brief visit. Other than a few architectural sketches—preliminary ideas concerning the proposed face-lift for Florence—Leonardo left nothing about this visit in his surviving notebooks.

The next stop was Bologna. There the pope met and conferred with the new French King, François I, only 21 years old and fresh from a victory over the Sforza and their Swiss mercenary allies. Brash and confident after retaking Milan,

François I knew of Leonardo from his predecessor, Louis XII, and was even more impressed and awestruck on meeting him in person.

A few months later, in March 1516, Giuliano de' Medici died and Leonardo was again without a formal protector and benefactor. Though no record survives of a formal invitation or summons from François, by the end of the summer Leonardo was heading for Amboise, France—the last and longest trip of his life. He was accompanied by Francesco Melzi and a servant. Though he would later make a visit to France, Salai most likely stopped off in Milan. In Amboise the king provided Leonardo with a spacious manor house, Manoir du Cloux, about a half mile away from his own château. Leonardo was 64 years old when he arrived in France and not in the best health.

In early 1517 Cardinal Luigi of Aragon, returning from a journey to Holland, visited Leonardo. The visit was recorded in the travel diary of the cardinal's secretary, Antonio de Beatis. Beatis notes that Leonardo no longer painted and that his right hand was paralyzed, probably from a stroke he suffered while in Rome. The notes relate that Leonardo showed the traveling party three paintings—the "Mona Lisa,"

*Jean Clouet, "Equestrian Portrait of François I,"
1524 Galleria degli Uffizi, Florence*

Following pages: The Château d'Amboise, in France's Loire Valley, was once the home of François I, Leonardo's last patron. Leonardo set up his household in the nearby Manoir du Cloux, known today as Le Clos Lucé.

the "Virgin and Child with St. Anne," and his final painting, "St. John the Baptist." (All are still in France, now hanging in the Louvre.) Beatis also remarks on Leonardo's anatomical drawings and on the fact that Leonardo told him that they were the result of performing 30 dissections of human corpses. Beatis praises Leonardo's accomplishments in anatomy and comments on some of the work found in other notebooks. The detailed discussion in Beatis's travel diary suggests that the notebooks and manuscripts must have been prominent and visible in the house; references to the notebooks in the diary also suggest that during these final years in France, Leonardo continued the never-to-be-finished project of preparing them for publication.

François-Guillaume Ménageot, "The Death of Leonardo da Vinci," 1781. Château d'Amboise, France

In May 1518, Leonardo staged his last theatrical extravaganza to celebrate the baptism of the king's son Henri and the marriage of the king's niece Madeleine to Lorenzo di Pietro de' Medici, Pope Leo X's nephew and the man in charge in Florence after the Medici returned to power. As usual, according to the surviving testimony, the crowds went wild. In June there was a party given at Leonardo's house in honor of the king, perhaps a last celebration of François and Leonardo's mutual regard. We hear little from Leonardo during the rest of 1518—his health, no doubt, in serious decline.

On April 23, 1519, Leonardo made a will leaving Francesco Melzi his books, paintings, notebooks, and manuscripts—essentially appointing him his literary executor. Melzi would turn in a mixed performance in this role. He put many of the papers in order following Leonardo's death, but he is also responsible for much of the scattering that occurred, readily lending materials that were often not returned, and he did not take sufficient care to insure that this great legacy would be properly cared for after his own passing. Leonardo left Salai the house and garden in Milan that had been restored to him after the return of the French to the city; to his half brothers, with whom he had been reconciled, he left his bank accounts in Florence.

Vasari tells a melodramatic, sentimental, and certainly apocryphal story of the death of Leonardo:

> [H]aving become old, [Leonardo] lay sick for many months, and finding himself near death and being sustained in the arms of his servants and friends, devoutly received the Holy Sacrament. He was then seized with a paroxysm, the forerunner of death, when King Francis I, who was accustomed frequently and affectionately to visit him, rose and supported his head to give him such assistance and to do him such favor as he could in the hope of alleviating his sufferings. The spirit of Leonardo da Vinci, which was most divine, conscious that he could attain to no greater honor, departed in the arms of the monarch.

Vasari goes on to say that Leonardo received instruction in the Catholic religion before being given his last rites, but this is also unlikely. Leonardo was certainly not a modern atheist, but the truth must be closer to what Vasari had said in the earlier 1550 version of the *Lives,* where he had called Leonardo "much more a philosopher than a Christian." Even the date of Leonardo's death, given by Vasari as May 2, 1519, is uncertain. On that date, as well as the surrounding days, other reliable evidence indicates that the king was elsewhere in the realm and not at Leonardo's bedside, though he might have been there if he had been available. It is all vintage Vasari. Simpler and straight to the point are the words of Francesco Melzi, who certainly was at Leonardo's deathbed: "May God Almighty grant him His eternal peace. Every one laments the loss of a man whose like Nature cannot produce a second time."

Leonardo was buried in the graveyard of the church of St. Florentin in Amboise. But his body fared no better than his literary remains. Like most church properties, St. Florentin suffered during the French Revolution, and in 1802 the church and graveyard were demolished, including the grave markers. The stone was used to rebuild the château up the road. The bones of the dead were reburied in one grave, without markers, by the church gardener. In 1863 poet and Leonardo admirer Arsene Houssaye excavated the church site and discovered a skeleton with an unusually large skull. He assumed he had found the remains of Leonardo. These bones were buried in the chapel of St. Hubert on the grounds of nearby Château d'Amboise. Even in death and beyond, in every way, Leonardo refuses to rest, refuses to be pinned down and defined.

"Every one laments the loss of a man whose like Nature cannot produce a second time."
—FRANCESCO MELZI

LEONARDO'S SKETCHBOOK

LEONARDO WAS A SCIENTIST DOING ART, and an artist doing science. Among all of his scientific endeavors—in physics, geology, meteorology, aeronautics, hydrology—his surpassing ability to wear both hats simultaneously is nowhere more effectively displayed than in his anatomical studies. This final gallery presents drawings from his studies of human and animal anatomy, from the human fetus to the embryo of a cow, with the pages all systematically annotated in his characteristic mirror script.

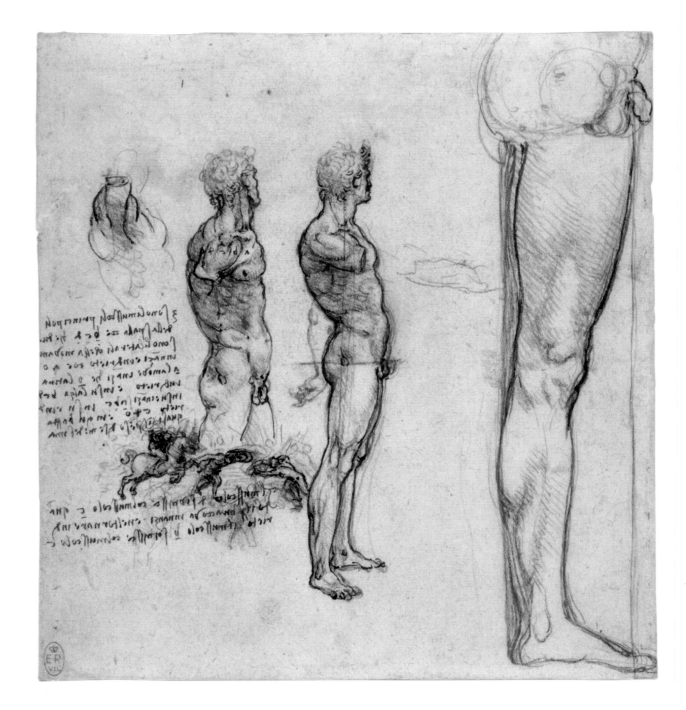

"Anatomical Drawings of a Male Nude, and a Combat Scene (lower left)," 1504-1506. Royal Collection, Windsor Castle, U.K.

"Studies of a Fetus in the Womb," ca 1511. Royal Collection, Windsor Castle, U.K.

"Rendering of the Embryo of a Cow," displaying the unmistakable shape of the logarithmic spiral, ca 1508. Royal Collection, Windsor Castle, U.K.

"Heart and Lungs of an Oxen," ca 1512-13. Royal Collection, Windsor Castle, U.K.

"Study of the Anatomy of the Neck," ca 1510-13. Royal Collection, Windsor Castle, U.K.

"Study for Human Musculature," 1508. Royal Collection, Windsor Castle, U.K.

> *"There may not be in the world an example of another genius so universal, so incapable of fulfillment, so full of yearning for the infinite, so naturally refined, so far ahead of his own century and the following centuries."*
>
> —HIPPOLYTE TAINE
> *VOYAGE EN ITALY*, 1866

EPILOGUE

In the centuries since his death, Leonardo's inventions and scientific discoveries have come to light long after so much that he discovered was rediscovered independently and long after entire fields that he invented were reinvented. Considering how much of the content of his notebooks was scattered about and lost, the questions and methods he introduced might well have been handed down by word of mouth, or through pages that were seen by others. But without evidence that these notes did get into the hands of others, we must assume that it was all rediscovered independently.

The Transformative Genius

The traditional view of the date for the birth of modern science is 1543, when two monumental books were published: the inspired treatise by Nicolaus Copernicus, *De Revolutionibus Orbium Coelestium (On the Revolution of the Heavenly Spheres)*, arguing on behalf of a sun-centered universe; and the first accurate anatomical atlas by Andreas Vesalius, *De Humani Corporis Fabrica (On the Fabric of the Human Body)*. By the 17th century, the Scientific Revolution was in full bloom with Galileo Galilei, who made fundamental discoveries in astronomy and classical mechanics; William Harvey, who demonstrated the circulation of blood; Robert Boyle, who discovered the relationship between pressure, volume, and temperature in a gas; Johannes Kepler, who formulated the laws governing the motion of the planets; and, most important, Isaac Newton, who produced a consistent theory of classical mechanics, co-invented calculus, discovered the law of universal gravitation, and made the connection between fundamental physical laws and the motions of heavenly bodies. Leonardo's manuscripts, however, offer convincing reasons to see his work as foreshadowing modern scientific methodology by decades before Copernicus and Vesalius produced their works. Although Leonardo never disseminated his discoveries by publishing them, compelling argument can be made that he was indeed the first modern scientist.

Opposite: St. Hubert's Chapel at Château d'Amboise, where, according to tradition, Leonardo is buried

It is easy to be awed by Leonardo's achievements—as well as intensely frustrated. His unfortunate reputation as someone who did not complete commissions was well deserved, resonating in Pope Leo X's dismissive refrain: "*Oime, costui non e per far nulla, da che comincia a pesare all fine innazi il principio dell'opera*" (Alas this man will never do anything, because he is already thinking of the end before he has even begun the work.) In truth, Leonardo was driven by an insatiable curiosity about the workings of nature, and when he was not experimenting with natural law, he was experimenting with all the facets of his artistic endeavors—from the composition to the medium. He was the consummate scientist, whether doing science or art. Giorgio Vasari, who obviously harbored immense admiration for

him, in describing the sublime quality and power of his art, clearly felt an obligation to explain that Leonardo's meager artistic output was due in large part to his distraction with science. But in writing that Leonardo "could have been a great scientist [if that was all he did]," Vasari revealed that he had little idea of the quality of Leonardo's scientific work.

We all know gifted individuals. They graduate at the top of their class, the smartest among them go on to the best colleges and universities and excel there. The extraordinarily intelligent among them might also make truly exceptional discoveries after concentrated and inspired study, and perhaps even win a Nobel Prize—though it would be a rare and immodest Nobel laureate who would not confess to a strong element of luck. And it would not be a stretch to label most Nobel Prize winners geniuses—but only "ordinary" geniuses!

Beyond the class of the ordinary geniuses there exists the rarefied class of "transformative" geniuses, comprised of the genuinely iconic figures in their

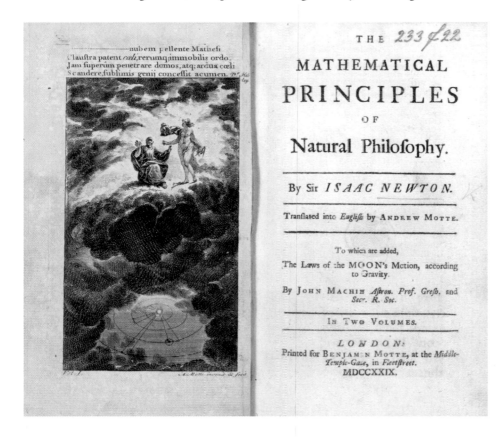

respective fields: Michelangelo, Beethoven (perhaps Bach and Mozart), Shakespeare, Newton, Archimedes, Einstein—individuals who invent, reinvent, or transform entire fields. In the sciences rarely would anyone knowledgeable in the field argue with the choice of Newton, or in literature, of Shakespeare, as a transformative genius. We have no idea whence they come, and there is no way to explain their gifts—which is part of their mystique.

Can there be physiological differences, mutations, or a particular concatenation of genes in individuals of this class? We may know one day, but we do not know now. In only one case is there a clue, but it could well be a red herring. Accordingly, it is safest to take it anecdotally. At Einstein's death, in 1955, Dr. Thomas Harvey, then a young pathologist at Mercer County Hospital in Princeton, was assigned to do Einstein's postmortem, after which he kept the great scientist's brain in a jar of formaldehyde. Decades later, when the news emerged about Einstein's brain having been preserved, Dr. Harvey, by then retired and living in Wichita, Kansas, retrieved the jar, hidden behind books on a bookshelf, and began to farm out slides—mounted slivers of Einstein's brain—to a number of groups performing neurophysiological research, as well as photographs of the brain in its entirety. One of the discoveries showed that Einstein's brain displayed a peculiar abnormality: It was lacking the parietal operculum, located above and behind the left ear—a key area for speech. The inferior parietal lobe, located just above the left ear, had grown about 15 percent beyond the normal size expected for Einstein's cranial cavity, and taken over the area of the missing operculum. Since spatial relationships and mathematical analyses are performed in the inferior parietal lobe, Einstein may have possessed a greater amount of the specialized gray matter required for mathematical work. In that sense, he may just have been hard-wired as a mathematical genius. But no other brains of individuals displaying the level of Einstein's intellect have ever been preserved, not Newton's, not Beethoven's, not Leonardo's.

Leonardo, who expressed insecurity as an *"omo sanza lettera"* (an unlettered man), demonstrated an asset far more noble: as an autodidact and as a *"disscepolo della sperientia"* (disciple of experience), the source of his learning was that ultimate primary source—nature itself. But other than that, what does he reveal about himself? As Newton biographer R. S. Westfall characterized his subject, one might also say about Leonardo: The deeper one attempts to penetrate his psyche, the farther he recedes into the mist. Sherwin Nuland, in a short but extraordinarily incisive biography, *Leonardo da Vinci* (2000), wrote a particularly stirring passage:

> If he is, as Sir Kenneth Clark so appropriately calls him, "the most relentlessly curious man in history," he is also the historical figure about whom we are most relentlessly curious.... The enigma of the "Mona Lisa's" smile is no less than the enigma of her creator's life force. Or perhaps the smile is in itself Leonardo's ultimate message to the ages: There is even more to me than you can ever capture; though I have spoken so intimately to you in my notebooks even as I have spoken to myself, I have kept final counsel only with the depths of my spirit and the inscrutable source that has made me possible; seek as you may, I will commune with you only so far; the rest is withheld, for it was my destiny to know things you will never know.

In art, Leonardo's works are few, too few even for a part-time artist, but they include the two most famous paintings in history, the most copied works in all of art. Thus even with an extremely modest output, he elevated art to a different level. And in a field in which quality counts far more than quantity, he is most certainly a transformative artist.

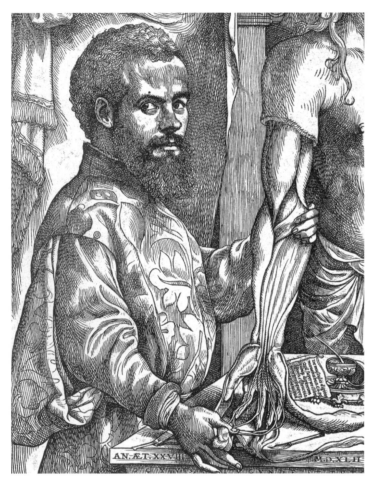

Above: Andreas Vesalius, seen in the frontispiece of his De Humani Corporis Fabrica, *published in 1543, when Vesalius was 29 years old*

Opposite: Godfrey Kneller, "Portrait of Sir Isaac Newton," 1689. Trinity College, Cambridge, U.K. The painting shows the physicist, astronomer, and mathematician at the peak of his celebrity.

As a scientist, Leonardo was practicing geology before the field was invented. His anatomical drawings have not been equaled. He made pronouncements on evolution. His studies in gravitation—the study of free fall, parabolic trajectories of projectiles—puts him under the same intellectual umbrella as Galileo, Newton, and Einstein. He designed a reflecting telescope. He invented robotics. He also invented more perfunctory devices—scissors, gears, springs, chain drives, the bicycle, topographic cartography, the odometer (to measure distance), the hygrometer (to measure humidity), the anemometer (to measure wind velocity), diving bells, and canal dredgers. He conceived the notion of the flying machine, parachute, and submarine.

He designed the machine gun, mortar, armored car, and other futuristic weapons. The list is endless. And we have only a quarter of his original notebooks. What other thoughts were in the missing three-quarters, now never to be seen? Although he prefigured myriad scientific and technological discoveries, he failed to publish them. If these had all been disseminated and taken to the next level by the generations of scientists who followed, we might have arrived at our level of scientific sophistication centuries earlier. Thus he is a clear failure as a transformative scientist.

Leonardo, like Newton, Einstein, and Mozart, remained a child at play throughout his life. Newton's words—"to myself I appear to have been like a boy playing upon the seashore and diverting myself now and then by finding a smoother stone or prettier shell than ordinary, while the great ocean of truth lay before me all undiscovered"— could easily have been uttered by Leonardo. And like Beethoven, he was a perfectionist, toiling on the minute details of his creations. It was never effortless, but rarely was it not perfect. His "Last Supper" is so great an achievement that it must be compared to supreme achievements in other fields—to Newton's *Principia*, Einstein's theory of general relativity, to Beethoven's Ninth Symphony—creations that elevate the human spirit and make it unimaginably richer.

Universal genius Leonardo, however, unlike the great scientists of the Scientific Revolution, is unique for the depth, breadth, and the quality of his mental inventions. If only one message is to be taken away from his modus operandi, it would be this: "It is important to be curious, and important to explore different intellectual worlds, but it is essential to seek their connections." On his deathbed in Amboise, he turned to his assistant, Melzi, and whispered, "Did anything get done?" Five centuries after his lifetime, we look back and wonder whether there weren't a dozen supremely gifted individuals all using the same name. Yet he was not expressing a false modesty when he wrote: "I have offended God and mankind because my work didn't reach the quality it should have."

FURTHER READING

The list of sources and suggested readings below is far from exhaustive. Leonardo studies have exploded in recent years, and new discoveries and insights follow one another with regularity. An excellent recent biography is Charles Nicholl's *Leonardo da Vinci: Flights of the Mind.* Nicholl is thorough, careful, and scholarly, yet accessible to the intelligent lay reader. On the subject of Leonardo's paintings, the definitive discussion of the artist's early period is found in David Alan Brown's *Leonardo da Vinci: Origins of a Genius.* For all the paintings, Pietro Marani's *Leonardo da Vinci* is comprehensive and authoritative. Carlo Pedretti's *Leonardo: Art and Science* presents the multifaceted nature of Leonardo's work. The anatomical studies are treated with professional reverence by Sherwin Nuland, professor of surgery at Yale, in his short biography, and by Kenneth Keele, physician and medical historian, in a series of scholarly papers. A highly useful source on the Internet for Leonardo's art is Maxine Anabell's writing on Lairweb. And to no one's surprise, the work that provides a unifying thread not just on Leonardo studies but also on the Italian Renaissance in general is Giorgio Vasari's still useful *Lives of the Painters, Sculptors and Architects.*

Ames-Lewis, Francis. *The Intellectual Life of the Early Renaissance Artist.* Yale University Press, 2000.

Atalay, Bulent. *Math and the Mona Lisa: The Art and Science of Leonardo da Vinci.* Smithsonian Books, 2004.

Bambach, Carmen, ed. *Leonardo da Vinci: Master Draftsman.* Exhibition catalog. The Metropolitan Museum of Art and Yale University Press, 2003.

Barcilon, Pinin Brambilla, and Pietro Marani. *Leonardo: The Last Supper.* University of Chicago Press, 1999.

Berktay, Halil. *Renaissance Italy and the Ottomans: History's Overlaps and Fault Lines.* Exhibition catalog. Sakip Sabanci Museum, Istanbul, 2003.

Bramly, Serge. *Leonardo: The Artist and the Man.* Penguin Group, 1995.

Bronowski, Jacob. *The Ascent of Man.* Little, Brown, 1973.

Brown, David Alan. *Leonardo da Vinci: Origins of a Genius.* Yale University Press, 1998.

Clark, Kenneth. *Civilization.* Harper and Row, 1969.

_____. *Leonardo da Vinci.* Penguin Paperbacks, 1989.

De Paoli, Dino. "Leonardo da Vinci's Geology and the Simultaneity of Time." *21st Century Science and Technology* (Summer 2001).

Fairbrother, Trevor, and Chiyo Ishikawa. *Leonardo Lives: The Codex Leicester and Leonardo da Vinci's Legacy of Art and Science.* University of Washington Press, 1997.

Goffen, Rona. *Renaissance Rivals: Michelangelo, Leonardo, Raphael, Titian.* Yale University Press, 2002.

Grendler, Paul. *The Universities of the Italian Renaissance.* Johns Hopkins University Press, 2002.

Hale, John R. *The Renaissance.* Time-Life Books, 1965.

Keele, Kenneth D. "Leonardo da Vinci and the Movement of the Heart." *Proceedings of the Royal Society of Medicine* (March 1951), 209–213.

_____. "Leonardo da Vinci's Influence on Renaissance Anatomy." *Medical History* (October 1964), 360–370.

_____. "Uses and Abuses of Medical History." *British Medical Journal* (November 19, 1966), 1251-54.

_____. "Leonardo's Anatomia Naturale." *Yale Journal of Biology and Medicine* (July-August 1979), 369–409.

Kemp, Martin, ed. *Leonardo on Painting,* Selected and trans. Martin Kemp and Margaret Walker. Yale University Press, 1989.

Kemp, Martin. *The Science of Art: Optical Themes in Western Art from Brunelleschi to Seurat.* Yale University Press, 1990.

King, Ross. *Brunelleschi's Dome: How a Renaissance Genius Reinvented Architecture.* Penguin, 2001.

_____. *Michelangelo and the Pope's Ceiling.* Walker and Company, 2003.

Kliner, Fred S., Christin J. Mamiya, and Richard G. Tansey, eds. *Gardner's Art Through the Ages.* 11th ed. Harcourt, 2001.

Landrus, Matthew. *The Treasures of Leonardo da Vinci.* HarperCollins Books, 2006.

Leader, Darian. *Stealing the Mona Lisa.* Shoemaker and Hoard, 2002.

Livingstone, Margaret. *Vision and Art: The Biology of Seeing.* Harry N. Abrams, 2002.

Marani, Pietro. *Leonardo da Vinci.* Harry N. Abrams, 2003.

Nicholl, Charles. *Leonardo da Vinci: Flights of the Mind.* Viking Penguin Group, 2004.

Nuland, Sherwin. *Leonardo da Vinci,* Penguin Lives, 2000.

Pater, Walter. *The Renaissance: Studies in Art and Poetry.* Oxford University Press, World Classics Paperback, 1986.

Pedretti, Carlo. *Leonardo: Art and Science.* Taj Books, 2004.

Pesco, Claudio, ed. *Leonardo Art and Science.* Giunti Gruppo, 2001.

Reti, Ladislao, ed. *The Unknown Leonardo.* McGraw-Hill, 1974.

Rigaud, John Francis, ed. *A Treatise on Painting by Leonardo da Vinci.* George Bell and Sons, 1897.

Sassoon, Donald. *Leonardo and the Mona Lisa Story: The History of a Painting Told in Pictures.* Madison Press Book, 2006.

Turner, Richard A. *Inventing Leonardo.* Alfred A. Knopf, Inc., 1992.

Vasari, Giorgio. *Lives of the Painters, Sculptors and Architects.* Everyman's Library, 1996.

Westfall, R. S. *Never at Rest: A Biography of Isaac Newton.* Cambridge University Press, 1981.

White, Michael. *Leonardo: The First Scientist.* St. Martin's Press, 2000.

Zöllner, Frank. *Leonardo da Vinci: The Complete Paintings (Vol. I)* and *Sketches and Drawings (Vol. II).* Taschen, 2004.

OTHER SOURCES

Bertelli, Carlo. "Restoration Reveals the Last Supper," *National Geographic* (November 1983), 664-685.

Glick, Daniel. "The Polish Leonardo," *Washington Post Magazine* (March 15, 1992), 16-20

Anabell, Maxine. www.lairweb.org.nz/leonardo/leda.html

ABOUT THE AUTHORS

BULENT ATALAY is a scientist and artist. Son of a Turkish military officer and diplomat, he was born in Ankara, Turkey, and educated in England and the United States. His early schools included Eton, U.K., and St. Andrew's School, Delaware. He received training in theoretical physics (B.S., M.S., M.A, Ph.D. and postdoctoral training) at Georgetown, Princeton, University of California–Berkeley, Oxford University, and the Institute for Advanced Study. He is a professor of physics at the University of Mary Washington and an adjunct professor at the University of Virginia. In art, he has shown in one-man exhibitions in London and Washington. In addition to technical papers, he is the author of *Math and the Mona Lisa: The Art and Science of Leonardo da Vinci* (Smithsonian Books, 2004), now translated in 11 languages. Dr. Atalay lectures around the world on a variety of subjects. www.bulentatalay.com

KEITH WAMSLEY was born in Seattle, Washington, and moved at the age of three to suburban New Jersey, where he graduated from public high school. After serving in the United States Navy he attended Cornell University, where he graduated in 1979 with a degree in classics. He received graduate training at Brown University and at the University of Mary Washington, where he received an M.A. He travels widely and currently lives in Fredericksburg, Virginia, where he teaches at a private secondary school. He served as copy editor for *Math and the Mona Lisa.* www.keithwamsley.com

ACKNOWLEDGMENTS

THIS BOOK IS THE RESULT OF CLOSE COLLABORATION BETWEEN THE AUTHORS AND A NUMBER of highly intelligent and talented individuals at National Geographic Books. It has ultimately proved to be a labor of love, demonstrating that working on any project dealing with Leonardo da Vinci creates "Leonardisti," or lovers and admirers of this extraordinary man who lived 500 years ago.

We are immensely grateful to Karen Kostyal, friend and a former editor at National Geographic Books, who first brought us into contact with Lisa Thomas, senior editor at National Geographic Books. We are also grateful to Kevin Mulroy, publisher, National Geographic Books, who has been highly supportive. Early in the project, Barbara Seeber served as the project manager while Lisa was on leave. Her enthusiastic support fueled our own enthusiasm. Dana Chivvis was the visionary photo editor, combining images of Italian Renaissance art with the National Geographic's own world-class photos. Art designers Sanaa Akkach and Cinda Rose are responsible for the design of the book that we believe would have impressed Leonardo himself. Copy editor Jane Sunderland demonstrated her remarkable editing skills and saved us from many errors, although any that survive are distinctly our own.

Away from the National Geographic Society, there were contributions by Sabrina Pezzoli, of Bologna, Italy, who in 2006 first brought to our attention images of Ponte Buriano near Arezzo, the bridge that Leonardo most likely used in the background of the portrait of Mona Lisa. Then in early 2008, Ms. Pezzoli introduced us to the Balze Rocks in Valdarno (also near Arezzo), which appear to have inspired the dreamlike vision serving as the general background of the "Mona Lisa." Stella Marinazzo, of Brindisi, Italy, served as a liaison with Marco Ramerini, whose photographs of the Balze Rocks we were able to include in Chapter Five, and also helped us iron out some lingering questions about medieval Italian. Ms. Marinazzo's son, Alex Medico, a graduate student at Harvard University in medieval history, provided helpful suggestions. We would like to express our gratitude to Kim Williams—architect, publisher, Renaissance scholar residing in Turin, Italy —who has been helpful in answering many questions.

Five hundred years ago, Leonardo offered his services to the sultan in Istanbul as an engineer to build an elegant single-arch bridge over the Golden Horn. Nothing came of Leonardo's design during his lifetime, but 25 generations later the imaginative Norwegian painter Vebjorn Sand spearheaded a project to build a scaled-down version of the bridge in Ås, Norway. The artist and his assistant, Melinda Iverson, provided photographs of the Leonardo-inspired span.

We are grateful to the talented painters Daniel Ludwig, Joseph DiBella, and Jeannine Atalay Harvey for fruitful discussions regarding the preparation of paint and gesso.

Leonardo harbored a lifelong passion to understand the human body, and his anatomical drawings are legendary, arguably the best ever produced. In a sidebar in Chapter Six, we discuss Leonardo's "Lessons for Cardiac Surgery." In this context, Dr. Elias Zerhouni, director of the National Institutes of Health, gave us access to rare medieval anatomical studies. These serve as counterpoints to Leonardo's own drawings. Celebrated British cardiac surgeon Dr. Francis Wells, who developed new techniques for repairing the mitral valve in the heart and was inspired by Leonardo's drawings and observations, was helpful in answering many questions. And for the modern CT-scan image of a heart, specifically to compare with Leonardo's own drawings, we are indebted to Dr. Michael Atalay of Brown University Medical School.

Finally, we are grateful to Carol Jean Atalay for many helpful discussions, and for calmly tolerating us during the frenetic pace we maintained in writing the book.

ILLUSTRATIONS CREDITS

INDEX **Boldface** indicates illustrations.

TEXT CREDITS:

Page 217 Reproduced by permission of the publisher from *Vision and Art: The Biology of Seeing* by Margaret S. Livingstone © 2002 Published by Harry N. Abrams, Inc., New York. All Rights reserved;

Page 276 From *Leonardo da Vinci* by Sherwin B. Nuland, copyright © 2000 by Sherman B. Nuland. Used by permission of Viking Penguin, a division of Penguin Group (USA) Inc.

LEONARDO'S UNIVERSE
The Renaissance World of Leonardo da Vinci
BULENT ATALAY AND KEITH WAMSLEY

PUBLISHED BY THE NATIONAL GEOGRAPHIC SOCIETY

John M. Fahey, Jr., President and Chief Executive Officer

Gilbert M. Grosvenor, Chairman of the Board

Tim T. Kelly, President, Global Media Group

John Q. Griffin, President, Publishing

Nina D. Hoffman, Executive Vice President;
    President, Book Publishing Group

PREPARED BY THE BOOK DIVISION

Kevin Mulroy, Senior Vice President and Publisher

Leah Bendavid-Val, Director of Photography Publishing
    and Illustrations

Marianne R. Koszorus, Director of Design

Barbara Brownell Grogan, Executive Editor

Elizabeth Newhouse, Director of Travel Publishing

Carl Mehler, Director of Maps

STAFF FOR THIS BOOK

Barbara Seeber, Lisa Thomas, Editors

Jane Sunderland, Text Editor

Dana Chivvis, Illustrations Editor

Cinda Rose, Art Director

Sanaa Akkach, Designer

Robert Waymouth, Illustrations Specialist

Jennifer A. Thornton, Managing Editor

R. Gary Colbert, Production Director

MANUFACTURING AND QUALITY MANAGEMENT

Christopher A. Liedel, Chief Financial Officer

Phillip L. Schlosser, Vice President

Chris Brown, Technical Director

Nicole Elliott, Manager

Monika D. Lynde, Manager

Rachel Faulise, Manager

Founded in 1888, the National Geographic Society is one of the largest nonprofit scientific and educational organizations in the world. It reaches more than 285 million people worldwide each month through its official journal, *National Geographic,* and its four other magazines; the National Geographic Channel; television documentaries; radio programs; films; books; videos and DVDs; maps; and interactive media. National Geographic has funded more than 8,000 scientific research projects and supports an education program combating geographic illiteracy.

For more information, please call 1-800-NGS LINE (647-5463) or write to the following address:

National Geographic Society
1145 17th Street N.W.
Washington, D.C. 20036-4688 U.S.A.

Visit us online at www.nationalgeographic.com/books

For information about special discounts for bulk purchases, please contact National Geographic Books Special Sales: ngspecsales@ngs.org

For rights or permissions inquiries, please contact National Geographic Books Subsidiary Rights: ngbookrights@ngs.org

Library of Congress Cataloging-in-Publication Data

Atalay, Bulent.
    Leonardo's universe : the Renaissance world of Leonardo
da Vinci / Bulent Atalay, Keith Wamsley.
        p. cm.
    Includes bibliographical references and index.
    ISBN 978-1-4262-0285-8 (trade) -- ISBN 978-1-4262-
0286-5 (deluxe)
    1. Leonardo, da Vinci, 1452-1519. 2. Artists--Italy--
Biography. 3. Inventors--Italy--Biography. 4. Renaissance--
Italy. I. Wamsley, Keith, 1951- II. Title.
    N6923.L33A885 2008
    709.2--dc22
    [B]
                                            2008010543

ISBN: 978-1-4262-0285-8 (regular)
ISBN: 978-1-4262-0286-5 (deluxe)

Printed in U.S.A.